# Contemporary Issues in Technology Education

**Series Editors**



P John Williams
University of Waikato, Hamilton, New Zealand

Alister Jones
University of Waikato, Hamilton, New Zealand

Cathy Buntting
University of Waikato, Hamilton, New Zealand

## Contemporary Issues in Technology Education - About this series

Technology education is a developing field, new issues keep arising and timely, relevant research is continually being conducted. The aim of this series is to draw on the latest research to focus on contemporary issues, create debate and push the boundaries in order to expand the field of technology education and explore new paradigms. Maybe more than any other subject, technology education has strong links with other learning areas, including the humanities and the sciences, and exploring these boundaries and the gaps between them will be a focus of this series. Much of the literature from other disciplines has applicability to technology education, and harnessing this diversity of research and ideas with a focus on technology will strengthen the field.

More information about this series at http://www.springer.com/series/13336

P John Williams • Alister Jones • Cathy Buntting
Editors

# The Future of Technology Education

Springer

*Editors*
P John Williams
Director
Technology, Environmental, Mathematics and Science Education Research Centre
University of Waikato
Hamilton, New Zealand

Alister Jones
Deputy Vice-Chancellor
University of Waikato
Hamilton, New Zealand

Cathy Buntting
Faculty of Education
University of Waikato
Hamilton, New Zealand

ISBN 978-981-10-1361-4 ISBN 978-981-287-170-1 (eBook)
DOI 10.1007/978-981-287-170-1
Springer Singapore Heidelberg New York Dordrecht London



Printed on acid-free paper

Springer is part of Springer Science+Business Media (www.springer.com)

# Preface

This book was born out of many conversations with colleagues around the world about the importance of mapping an agenda for technology education moving forwards. One seed was planted about 10 years ago at a conference in Australia where John Williams and Kay Stables had a discussion about the merits of facilitating a symposium of technology educators who could discuss and debate future directions of technology education. More grounded thinking was done when working on the *International Handbook of Research and Development in Technology Education*, published in 2009. This book highlighted some of the significant advances that have occurred in technology education as a curriculum area with its own identity and value. For example, there is now much stronger alignment between technology education as a field and the history and philosophy of technology more generally, and an increasing focus on research in teacher development and student learning. However, the book also pointed out the necessity of continued political engagement as well as meaningful classroom-based research if the field is to continue to move forward.

Reflecting on the process of developing the Handbook in a paper published by the *International Journal of Technology and Design Education* ('The developing field of technology education: A review to look forward'), Alister Jones, Cathy Buntting and Marc de Vries proposed that an international meeting be held to articulate an agenda for ongoing research and development.

This book—*The Future of Technology Education*—is a result of these developments. Technology education researchers, many of whom are internationally renowned for their contributions to the field, were invited to contribute to the project. In order to create a cohesive contribution to the literature, while still enabling authors to assert their own voices, we first met together at a workshop in Stockholm in June 2012 to review chapter summaries and engage in formative discussions about each of the chapters. A subsequent meeting of all authors took place in March 2013 at Teachers College Columbia in New York to critique the full chapter drafts, which were subsequently updated to reflect the group's conversations at the workshop.

The significance of meeting at the place where John Dewey spent over two decades as Professor of Philosophy was not lost on any of us, and in Chap. 9 of this book David Spendlove reminds us of Dewey's (1916) caution:

> Nothing has brought pedagogical theory into greater disrepute than the belief that it is identified with handing out to teachers recipes and models in teaching.

This book is neither a recipe, nor a model. Rather, it brings together the informed musings of academics grappling with issues that we believe are likely to be of key importance in the future of technology education as a school subject. While as individuals or a collective we obviously do not propose to be foretelling the future, what we offer is a synthesis of issues that we believe should not be ignored.

It has been a privilege to work on this project with colleagues who are each committed to the ongoing development of the field of technology education. We trust that our collective thinking will contribute meaningfully to future directions.

Hamilton, New Zealand

P John Williams
Alister Jones
Cathy Buntting

# Contents

# About the Authors

**Frank Banks** is Emeritus Professor and the former Director for International Development in Teacher Education at The Open University in the United Kingdom. In that post he held responsibility for teacher professional development projects in Bangladesh, India and Sub-Saharan Africa. He also directed the innovative online initial teacher education programme for the University. Before joining the Open University he worked as a school teacher of science, engineering and technology in secondary schools in England and in Wales, including head of department, and as an elementary school advisory teacher in Powys, Wales. Frank has been a consultant in the professional development of teachers to Argentinean, Egyptian, South African and UK government agencies, UNESCO and the World Bank.

**David Barlex** is an acknowledged leader in design and technology education, curriculum design and curriculum materials development. He taught in comprehensive schools for 15 years, achieving head of faculty positions in science and design and technology before taking university positions in teacher education. He directed the Nuffield Design and Technology Project and was Educational Manager for Young Foresight. David is well known for his interest and expertise in developing curriculum materials that support pupil learning from a constructivist perspective. He uses this approach to develop young peoples' ability to understand and critique the design decisions made by professional designers and those they make themselves in design and technology lessons. This informed the Nuffield Design and Technology publications, which have been widely used in the UK and emulated abroad in countries such as Russia, Sweden, Canada, South Africa, Australia and New Zealand. David is semi-retired and lives with his wife Marion Rutland in Putney, West London.

**Cathy Buntting** has a background in biochemistry and biotechnology research and teaching and a Ph.D. in education. She holds a senior research position within the University of Waikato, New Zealand, and currently manages the internationally recognised Biotechnology and Science Learning Hubs. Her research interests focus on how digital technologies can transform science and technology teaching and learning, and on the development of science and technology education policy. She was Associate Editor of the *International Handbook on Research and Development in Technology Education.*

**Vanwyk K.M. Chikasanda** is a 2012 Ph.D. graduate of the University of Waikato, New Zealand, and working at the University of Malawi–The Polytechnic as Dean of the Faculty of Education and Media Studies and Lecturer in the Department of Technical Education. Vanwyk is currently researching pedagogical practices in science and technology education in Malawi primary schools. He also participated in the secondary school curriculum review as an expert in metalwork, which is offered in secondary schools. The review incorporated design as a component in the subject despite the drive to maintain the craft nature of the curriculum. Vanwyk is also coordinating a European Union funded scoping study of the education sector focusing on marketable skills development for Malawi.

**Marc J. de Vries** is Professor of Science and Technology Education and Affiliate Professor of Christian Philosophy (of Technology) at Delft University of Technology and Assistant Professor of Philosophy of Technology at Eindhoven University of Technology, both in the Netherlands. He is the current editor-in-chief of the *International Journal of Technology and Design Education* (Springer) and editor of the *International Technology Education Studies* book series (Sense). He is chairman of the board of the PATT Foundation that initiated the international PATT conferences and a member of the ITEEA Academy of Fellows. His numerous publications include a book on the philosophy of technology for technology educators (*Teaching About Technology*) and a textbook series for technology education for Dutch secondary schools (*Technologisch*).

**Marilyn Fleer** (PhD, MEd, MA, BEd) holds the Foundation Chair of Early Childhood Education at Monash University, Australia, and is the President of the International Society for Cultural Activity Research (ISCAR). Her research interests focus on early years learning and development, with special attention on pedagogy, culture, science and design and technology.

**Wendy Fox-Turnbull's** research interests include authentic learning in technology education, the role and nature of effective conversation in learning, particularly in technology, teaching and learning approaches for the twenty-first century, and assessment in technology education. A Senior Lecturer at University of Canterbury's College of Education, Wendy lectures in Technology Education and Professional Education Studies, Inquiry Learning and Teaching, and Assessment. Wendy is the current Chair of Technology Education New Zealand.

**Mishack T. Gumbo** is an Associate Professor in the Department of Science and Technology Education in the College of Education, University of South Africa. He has a Ph.D. with specialisation in indigenous technology as part of indigenous knowledge systems. He currently supervises masters and doctoral students, mainly in the areas of indigenous technology, technology education (with special reference to curriculum and teaching, teacher development, and pedagogical content knowledge), educational technology and multicultural education. Current and recent research projects include technology teachers' pedagogical content knowledge and postgraduate supervisors' use of ICTs.

**Alister Jones** is a Research Professor and Deputy Vice-Chancellor of the University of Waikato. Prior to this he was Dean of Education and the Director of the Wilf Malcolm Institute of Educational Research and the Centre for Science and Technology Education Research. He contributed significantly to the development of the technology curriculum in New Zealand, and has published extensively in technology and science education as well as general education. He is Director of a number of companies, including Cognition Education Limited, and Managing Director of the Australasian Science Education Research Association Limited.

**Steve Keirl** is Reader in Design Education at Goldsmiths, University of London, contributing to undergraduate and postgraduate Design and Design and Technology (D&T) Education programmes. His career spans over 30 years in D&T primary, secondary and tertiary education in Australia and England. In Australia, Steve chaired the Technology Expert Working Group and was principal author of the D&T section of the South Australian Curriculum Framework. He also contributed to the Investigation into the Status of Technology Education in Australian Schools and the OECD Investigation into Technology Education in Australia. Steve's research interests include the relationships among ethics, democracy, curriculum, pedagogy, design and technology—sometimes summarised as 'explorations of technological and design literacy'. He collects washing-up brushes.

**David Mioduser** is Professor of Science and Technology Education in Tel Aviv University's School of Education. He heads the Science and Technology Education Center and the Knowledge Technologies Lab. His research interests focus on cognitive and learning issues in technology education, particularly with young children, and the meaningful incorporation of advanced technologies into teaching and learning processes. His research has contributed to the development of innovative learning environments, a holistic technological-thinking model implemented in kindergartens in Israel, as well as curricula and learning materials for the elementary and high-school levels in science and technology.

**David Spendlove** is Deputy Director of Teacher Education at the Manchester Institute of Education at The University of Manchester. He was previously Head of Technology and a senior teacher before moving into Higher Education in 2000. His research has investigated the underachievement of boys in design and technology, creativity and emotion in design, and most recently design thinking. In addition, David has written books on emotional literacy, design and technology, and assessment for learning as well as reports on STEM in D&T and advanced level design and technology. He currently co-edits *Design and Technology Education: An International Journal*. In 2013 David received an Outstanding Contribution Award from the Design and Technology Association.

**Kay Stables** is Co-Director of the Technology Education Research Unit (TERU) at Goldsmiths, University of London. She began her career as a textiles teacher, joining Goldsmiths in 1986 as a researcher on the Assessment Performance Unit's D&T project. A founder member of TERU, she has directed and contributed to projects in primary and secondary education in the UK and overseas. With Richard Kimbell, she authored the TERU retrospective: *Research Design Learning* (2007). Recent research has focused on design, creativity and sustainable development, the use of digital tools in assessment (the e-scape project) and designerly well-being. She takes a keen interest in international design and technology education developments and has been a visiting scholar at institutions in Australia Canada, New Zealand, Sweden and the USA.

**Kendall N. Starkweather** is Executive Director/CEO Emeritus of the International Technology and Engineering Educators Association (ITEEA) and the Foundation for Technology and Engineering Education (FTEE). His career has focused on advancing technology education worldwide as a teacher, teacher educator and association executive. His experiences have included the development and implementation of legislation, promoting technology education in and outside of the educational community, advancing contemporary curriculum and instruction, working on core competencies through standards and certification processes, and positioning and raising the value of technology education through quality educational practices. He has been recognised as a distinguished graduate from the University of Maryland where he earned his Ph.D., a Distinguished Technology and Engineering Educator (DTEE) and member of the Academy of Fellows from the ITEEA, and a Certified Association Executive (CAE) from the American Society for Association Executives.

**P John Williams** is a Professor and Director of the Technology, Environmental, Mathematics and Science Education Research Centre at the University of Waikato in New Zealand, where he teaches and supervises research students in technology education. Apart from New Zealand, he has worked and studied in a number of African and Indian Ocean countries and in Australia and the USA, including directing the nationally funded *Investigation into the Status of Technology Education in Australian Schools*. His current research interests include mentoring beginning teachers, PCK, and electronic assessment of performance. He regularly consults on technology education in a number of countries, and is a longstanding member of eight professional associations. In 2011 he was elected to the International Technology and Engineering Education Association's Academy of Fellows for prominence in the profession.

# Chapter 1
# The More Things Change, the More (Some) Things Stay the Same

**Cathy Buntting, P John Williams, and Alister Jones**

The world is changing at an unprecedented rate—technologically, economically, environmentally, socially. In this 'knowledge age', where knowledge is the 'new currency', students require skills to access, analyse and evaluate information that is constantly changing, and use—and contribute to—this changing information, in collaboration with others, in order to support decision making, development and innovation.

Within this context, technology education has an increasingly important role to play in the school curriculum. The subject Technology Education (variously called Design and Technology, Technology, and Engineering) can provide students with opportunities to integrate their technological, economic, environmental, and social worlds, and develop their technological literacy. For example, by designing, developing and critiquing technological artefacts, processes, and systems, they can work with others and develop a range of cognitive and manipulative skills appropriate for the 'knowledge age'.

This book is offered as a platform from which to continue discussions about how technology education might progress into the future. Written by leading academics in the field of technology education, it takes as its starting point the assumption that, even in an ever-changing environment, some things stay the same—the critical importance of a sound curriculum that meets the needs of students, teachers and the

C. Buntting (✉)
Faculty of Education, University of Waikato, Hamilton, New Zealand
e-mail: buntting@waikato.ac.nz

PJ. Williams
Technology, Environmental, Mathematics and Science Education Research Centre, University of Waikato, Hamilton, New Zealand
e-mail: pj.williams@waikato.ac.nz

A. Jones
Deputy Vice-Chancellor, University of Waikato, Hamilton, New Zealand
e-mail: a.jones@waikato.ac.nz

 
PJ. Williams et al. (eds.), *The Future of Technology Education*, Contemporary Issues in Technology Education, DOI 10.1007/978-981-287-170-1_1

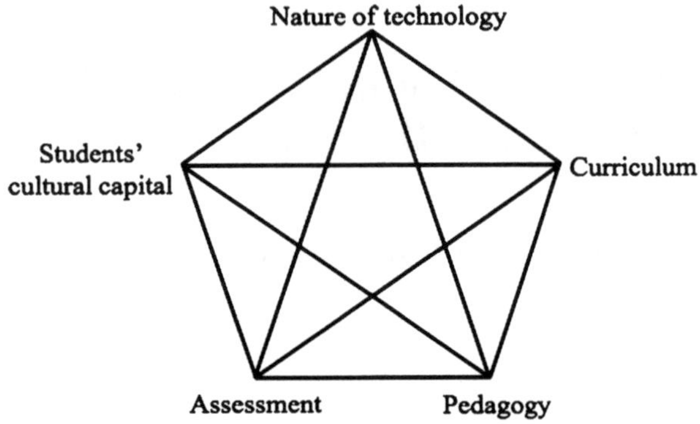


**Fig. 1.1** Key variables—and the interactions between them—that drive the technology education experienced by students

community, for example, and the need for national and state policy that coherently connects government and community aims across all levels of education. Although the detail within curriculum, assessment and surrounding education policy might change, the basic principles do not.

One way of conceptualising this is to consider the following as key variables driving the technology education that students experience in schools: curriculum, pedagogy, assessment, how the nature of technology is interpreted, and how students' cultural ways of knowing and acting (their cultural capital) are understood. If each of these is represented as the points of a pentagon (see Fig. 1.1), it is the interactions between them that become important in the provision of a meaningful form of technology education.

Importantly, it is teacher beliefs and values related to each of these variables that ultimately influence what happens in the technology education classroom. For example, a teacher's understandings of and beliefs about the nature of technology influence how he or she interprets curriculum documentation and what he or she chooses to emphasise in the classroom. But this is not a unidirectional interaction—the teacher's reading of the curriculum has potential to influence his or her understanding of the nature of technology. However, teacher perceptions of the nature of technology are often strongly embedded, and changes in the curriculum do not translate easily into changes in teachers' understandings of the nature of technology. Rather, sustained professional learning opportunities are needed. This is particularly important when new elements are introduced to the curriculum, such as the nature of technology and emphases on technological literacy.

The teacher's understandings of the nature of technology, interacting with understandings of the curriculum, will impact on the pedagogical approaches that are adopted. Thus, a view that technology education is about 'design and make' will likely drive classroom interactions that focus on these sorts of tasks and activities, rather than on developing students' understandings of other aspects of the nature of technology, such as sociocultural considerations. Again, the interactions are multi-directional. For example, teachers who have a preference for pedagogies that involve project work may spend a larger proportion of class time focusing on curriculum outcomes that can be achieved using this type of pedagogical approach, with other intended learning outcomes accorded less attention.

Of course, many teachers take as their starting point the curriculum (and their interpretations of it), develop learning outcomes relevant to their students, and use a range of appropriate pedagogies to address the varied learning outcomes—altering classroom approaches based on feedback from formative and summative assessment processes. Indeed, expert teachers are continually making decisions about learning outcomes, the pedagogical approaches best suited to support students to achieve these outcomes, and aspects of assessment—how to assess, when to assess, and the purpose(s) of the assessment (for accountability purposes, to give students feedback on their learning, to help students progress in their learning, to indicate when a new pedagogical approach might be needed, to offer feedback on students' progress to their parents, etc.).

Curriculum, pedagogy and assessment therefore interact in an on-going and fluid way throughout the teaching and learning interaction—guided, as indicated above, by the teacher's understanding of the nature of technology and his or her beliefs about what the primary purpose(s) of technology education might be. Another key variable driving the learning experiences of school students is the teacher's understandings of students' cultural capital—how students learn and act in their cultural context. The relevance of students' cultural capital is being increasingly recognised and articulated in both academic and teacher communities. For example, sociocultural and cultural-historical learning theories specifically take into account the cultural context in which learning takes place—where culture refers to far more than ethnic grouping, but groups with a commonality in understanding about certain aspects of the world. Here, a number of questions arise:

- Are the diverse knowledges and experiences that student bring with them to the classroom acknowledged, valued and incorporated into the learning—or are they ignored and so de-valued?
- What are the implications of either of these approaches for pedagogical decision making?
- How might a teacher's beliefs about pedagogy and assessment influence their understandings of students' cultural capital, and vice versa?
- Does the curriculum (both intended and implemented) create room for cultural capital to be explored, or is it highly prescriptive and geared towards the presumed likely out-of-school experiences of a dominant cultural group?
- What about the nature of technology—is technology seen as being culturally embedded?
- What learning characteristics run across cultures, and what are particular to certain cultural groups?

These questions, and the examples before them, highlight the complex, dynamic nature of teaching and learning in the technology education classroom. One of the significant factors not shown in Fig. 1.1, but wielding significant influence, is the role of politics in shaping how these variables interact—both at the macro (national or state) level and micro (school and classroom) level. For example, politics drive the development of the intended curriculum, influence the level of professional development, dictate forms of assessment and guide funding provisions. These are significant determinants of the nature of technology education, and perhaps can be considered to balance (but sometimes distort!) the shape of the pentagon in Fig. 1.1. For example, an over-emphasis on the accountability function of assessment can lead to the development of tasks that are easy to administer and evaluate but that may not align with aspects of the intended curriculum that are more difficult to assess. This can lead to distorted interpretations of what is important in the curriculum, what is taught, how it is taught, how (and even if) students' cultural capital is incorporated, and so on. Likewise, political drivers emphasising vocational outcomes over more general technological literacy outcomes (and vice versa) influence the ways in which different aspects of the nature of technology are explicated in curriculum documents, pedagogy, assessment, and the valuing of students' cultural ways of knowing and acting.

Research is another powerful variable not included in Fig. 1.1, but when focused on different parts of the pentagon can help us to understand the variables and their interactions. It should also contribute to the on-going development of technology education. This book, therefore, offers a salutary call for research that is able to inform future developments in technology education, and that is accessible to teachers and informed by their interests and needs. Teachers, as the gatekeepers through whom all the variables come together to determine what is experienced in classrooms, need to be critically cognisant of each of these variables in order to make the best decisions possible for their students' learning in technology education.

This book—intended for teachers, pre-service teachers, teacher educators, policy makers, and technology education researchers—aims to support discussions and decision making about how technology education might progress into the future. The framework presented in Fig. 1.1, with additional consideration of politics and research, is offered as a mental organiser generally, and as a conceptual structure for this book. A summary of each of the chapters and their contributions to this framework is offered below.

Building on the notion that understandings and beliefs about the nature of technology influence teachers' implementation of curriculum, pedagogy and assessment, Steve Keirl in Chap. 2, ***'Seeing' and 'interpreting' the human-technology phenomenon***, helps open this book by formulating and addressing the two-pronged challenge of how to understand technology, and how this might inform the development of technology education in the future. The philosophical study of technology began comparatively recently, partly because other academic areas, such as history and sociology, assumed this territory. Steve's approach is to use hermeneutics to analyse technology as binaries, such as human and technology, visible and invisible, positivist and antipositivist, utopian and dystopian, subject and object, and democratic and non-democratic.

Importantly, these binaries are not 'either-or', but enable analysis across a spectrum. In addition, a hermeneutics approach not only explores technology holistically and analytically, but also opens space for considerations of cultural, historical and political relationships. Considering technology as a series of 'both-at-once' binaries therefore enables its complexity to become manageable.

This logic can also be applied to technology education, where binaries could include: status quo and change agent, local and global, traditional and emergent, product and process, technical and designerly, and academic and practical. Such a binarial approach offers a way of interrogating the rich and complex areas of technology and technology education, to develop informed ways forward.

Following on from Steve's chapter, Marilyn Fleer outlines in Chap. 3, ***Theorising technology education from a cultural-historical perspective***, why one of the other key variables driving the technology education experienced by school students is teachers' understandings of students' cultural capital—their cultural ways of knowing and acting. Marilyn draws on cultural-historical theory in order to theorise a futures orientation for technology education, discussing the three interrelated ideas of tools and signs as cultural practice, everyday concepts and technological concepts, and imagination and creativity. Bringing these together in technology education offers a way for students and teachers to together consider the way the cultural and historical (and future) ecology of societies is realised through the knowledge base and values-oriented activity of design and technology, and to understand that technological activity is embedded in a deep knowledge of self and community.

Technology education develops the individual to imagine and create things not yet designed and produced. Through this, technology education contributes to the technological development of society. As such, Marilyn argues that technology education has both an ethical and moral responsibility to support imaginings that sustain people and communities in harmony and for the well being of the broader ecological and social environment. In projecting to the future, technology education must be conceptualised both pedagogically and morally as a social and ethical practice.

The importance of community knowledge in technological endeavours is picked up again by Mishack T. Gumbo in Chap. 4, ***Indigenous technology in technology education curricula and teaching***, in which he develops the case for integrating indigenous technology in technology education. The premise of this chapter is that indigenous technologies have an important place in technology education, although these are recognised as a gap in technology education curricula world-wide. Adopting a social-cultural-historical perspective, Mishack explains the predominance of Western Knowledge Systems in terms of post-colonial theory, exacerbated by the polarisation of the 'western' and 'southern' worlds. He calls for the integration of indigenous technology into technology education curricula to enhance the collective and progressive participation of the custodians of both Western Knowledge Systems, and Indigenous Knowledge Systems—where 'indigenous knowledge' is defined as the complex set of activities, values, beliefs and practices that have evolved cumulatively over time, and is active among communities and groups who are its practitioners. These communities may be national (e.g., Canadian Eskimos or New Zealand Māori) or task defined (e.g., minaret builders in Yemen or fine-woodworkers in London). The focus in this chapter is on national indigeneity, with Mishack drawing extensively on the South African education context.

As Mishack highlights, if the technology education curriculum is to be culturally sensitive, it must integrate indigenous technology. Students of the dominant culture are harmed by a curriculum that represents, affirms, and celebrates only their cultural background and experience. Students whose cultural backgrounds are not validated by the curriculum receive the implicit message that their cultures are not worthy of study, and that people of these cultures have achieved little. While this chapter is likely to be read with interest by those in countries with a recognised indigenous population, it has important messages for all who are interested in technology education, whether as teachers, researchers or administrators. Very few countries can currently claim monocultural status, with the vast majority grappling with how to enhance education in the culturally diverse classrooms that have resulted from globalisation.

Together, Chaps. 3 and 4 demonstrate how teachers' understandings of students' cultural capital influence the pedagogies that they adopt—and the empowerment that can be afforded by recognising and embracing students' cultural ways of knowing. They also impact on teachers' understandings of the nature of technology (as a cultural activity), curriculum (which needs to be culturally sensitive in order for diverse students to engage), and assessment (which will also need to value students' cultural capital and ways of knowing if these are to be truly valued in the teaching and learning).

In Chap. 5, ***The pedagogical ecology of technology education: an agenda for future research and development***, David Mioduser weighs up the need to consider what we already know about effective pedagogical models for technology education with what might be required of technology education in the future, and how these goals might be addressed. To take into account the radical changes in the social, economic and cultural realities of the twenty-first century, including technological transformations, a thorough revision of teaching and pedagogical approaches is necessitated. To pursue this agenda, David proposes a research and development pedagogical ecology framework that includes conceptual, contextual, pedagogical resources, and planning and implementation dimensions.

There is ample consensus that engagement in doing technology is an essential requirement for meaningful learning in technology education, but the pedagogical challenge lies in creating authentic learning contexts that can be holistically experienced by learners as credible and immersive, during which they develop conceptual as well as artefact-embedded technological knowledge. Teaching and learning in technology education also needs to be viewed as a complex system comprising many interacting factors affecting both individual and social learning processes. For example, the integration of mobile devices with cloud technologies and ubiquitous wireless communication both requires and facilitates researchers and educators to explore innovative pedagogical models and conceptualisations of technology-supported teaching and learning. This is picked up again in the final chapter by Marc J. de Vries, who considers the role of research in the ongoing development of technology education as a learning area that meets the needs of current and future students and their communities.

Building from David Mioduser's more general consideration of pedagogies for technology education, Wendy Fox-Turnbull focuses in Chap. 6, ***Conversations to support learning in technology***, on one aspect of pedagogy—the importance of 'talk' in the learning process. Using sociocultural theory as a framework, she highlights the significant role talk plays in cognitive development through group interactions and collaborative learning. Three major and interrelated themes in technology education conversations are introduced: deployment, conduit and knowledge. Deployment relates to the deployment of students' existing and recently learned knowledge, either from home, community or culture, or knowledge learned at school. The conduit theme refers to talk that supports the deployment of knowledge, skills and experiences into the learning context. Finally, the knowledge theme represents conversations that result from deployment and conduit—conversations describing the technological knowledge and skills gained through technological practice.

Notions of 'inquiry learning' and 'twenty-first century learning' place the student at the centre of the learning process, and when implemented together with informed understandings of the importance and nature of conversation, facilitate deep learning. Building from this, Wendy argues that technology education is powerfully positioned to develop in students competencies related to inquiry and life-long learning. 'Authenticity' in learning and learning tasks is paramount here, as is the clear articulation of appropriate learning intentions and how these will be pursued. Linking back to Chap. 3 by Marilyn Fleer, Wendy also reminds readers of the significance of the funds of knowledge that students bring to the classroom. Specifically, she points out the valuable learning that can be achieved when students' funds of knowledge are coupled with intercognitive conversation.

Kay Stables begins Chap. 7, ***Assessment: feedback from our pasts, feedforward for our futures***, with the reminder that the purpose of assessment determines its approach: assessment may be *of* learning or *for* learning. Kay goes on to point out that a behaviourist paradigm still largely dominates assessment practices, but this is in conflict with more progressive understandings of constructivist and sociocultural approaches to learning and teaching. In addition, different stakeholders have different positions on learning, increasing the complexity of the assessment context. This complexity is increased still further when considering how different stakeholders perceive feedback from assessments, and the potential consequent actions (the feedforwards) that are conceived.

In order to progress assessment in technology education, Kay focuses on four aspects: adopting a pedagogic approach to assessment (where both teaching *and* assessment practices aim to support learning), maintaining authenticity in activities through which assessment is being undertaken, recognising the importance of judgement in valid processes of assessment, and maintaining a focus on equity and the inclusive role of the learner. Kay also considers the potential affordances of new technologies in assessment. Harking back to the interactions represented in Fig. 1.1, She concludes the chapter by pointing out that future developments in technology education should support teachers to align pedagogy with assessment, while taking into account the cultural capital of students in authentic activities so that they can make visible their developing technological understanding and capability (the nature of technology in Fig. 1.1) to themselves and their teachers and assessors.

In Chap. 8, ***Developing a technology curriculum***, David Barlex also brings many of the variables presented in Fig. 1.1 together by considering how curriculum development and innovation is influenced by understandings of the nature of technology, what students might learn, how they might learn, and what the primary purpose(s) are for engaging in technology education. Within this frame, David proposes four main arguments for developing a technology curriculum—economic, utility, democratic and cultural—pointing out that there is a relationship between the argument and the type of curriculum that results. In addition, curriculum development is generally initiated by politicians, in collaboration with those who have the expertise, but it is rarely a straightforward process. The political desire for a product that can be used to show progress, and educators' recognition that technological knowledge is socially constructed, results in a tension.

As a way of informing technology curriculum development, David proposes three procedural principles: being true to the nature of technology, developing a perspective on technology, and enabling technological capability. Through explicating each of these principles and considering how they might play out in future developments for technology education, David is able to argue that using these principles will result in a curriculum that is a valid and worthwhile endeavour for all students, and that will facilitate the introduction of new elements, enabling the curriculum to keep pace with changes outside of school. Importantly, this chapter offers scope for the reader to consider where they stand with respect to the purpose (s) of technology education, and what this might mean for their understandings of the other variables—the nature of technology, the cultural capital of students, pedagogy and assessment.

David Spendlove approaches Chap. 9, ***Developing a deeper understanding of design in technology education***, from the position that design is central to being human; everyone designs and engages in the process of designing. However, while designing is an innate capacity, it is also a disciplined activity system located in industry, commerce, the arts and education, and has multiple definitions and uses. In other words, design is different in different contexts. Common across contexts is that all design should be creative, although not all creativity involves design. Design also involves riskiness and uncertainty, and is an integral aspect of a sustainable economy, ethical lifestyle and the shaping of communities.

Unfortunately, areas such as creativity, riskiness and uncertainty have become increasingly marginalised in educational contexts demanding ever-greater accountability in terms of productivity and performativity. In addition, undemanding and/or preconceived tasks can give the appearance of students designing and progressing when the reality is that learners lack autonomy or creative opportunity. Design is an integral aspect of a sustainable economy, ethical lifestyle and the shaping of communities, and David warns against a 'sterile form of design education' that fails to recognise social, psychological, philosophical, political, economic and cultural influences, or how design and technology have interacted in a free-market economy to shape contemporary culture. Rather, he argues for future-focussed technology education programmes that value ‘design thinking’ and how this can contribute to students’ learning and ‘being’.

In Chap. 10, ***The alignment of technology with other school subjects***, Cathy Buntting and Alister Jones examine the traditional separation, or siloing, of knowledge domains into distinct school subjects—languages, mathematics, science, social science, the arts, technology—and consider the benefits and challenges of breaking down these silos and bringing about closer alignment between technology as a school subject, and other subject areas. One key purpose for aligning technology with other subjects is to increase the scope and opportunities for students to develop relevant skills and dispositions to address ‘wicked problems’—complex problems with multiple causes and interdependencies that are difficult or even impossible to solve, or even define, using the tools and techniques of only one organisation or discipline.

Teacher knowledge is key if technology learning outcomes are to be identified, pursued, and assessed. This of course applies to school programmes where technology is taught as a separate ‘subject’, but it is perhaps even more important in courses where technology is aligned—or integrated—with other subjects. Without deeply embedded understanding of the learning goals for technology education, and how to achieve these, the integration of technology with other school subjects (often science) will likely limit the breadth and depth of technology learning that is achieved. However, an increasing range of examples is beginning to emerge demonstrating how technology learning can be integrated with other learning goals, and the broad range of outcomes that this has for students.

Just as the relationship between technology education and other school subjects needs ongoing consideration, the differences and overlaps between vocational and general technology education continue to need to be addressed. As John Williams outlines in Chap. 11, ***Vocational and general technology education***, the divide between vocational and general goals is not always clear. Further, the generalisation of vocational education and the vocationalisation of general education often results in a confluence of their goals. As John points out, the content (general vs. specific), goals (general vs. vocational), pedagogy (student-centred vs. other-centred) and assessment (outcomes vs. competencies) of vocational and general approaches are different.

Nonetheless, achievement of the goals of technology education—whether vocational or general, or both—is dependent on effective transfer of learning. John argues that instead of transfer being viewed as the re-application of skills and knowledge from one context to another, it should be seen as a process of boundary crossing that involves consequential transitions in which learners are engaged in a variety of quite different social, intellectual and manipulative tasks in different contexts. Once again, curriculum, pedagogy, assessment, and understandings of the nature of technology all interact in complex and dynamic ways. John predicts that, in the future, vocational programmes will become more student-oriented as a broader range of subjects become available for selection across both vocational and general technology areas. He also posits that because proficiency in technical and para-professional skills will not be enough in most work places, technology education will increasingly focus on more generic skills.

Somewhat in contrast to John William's discussion about the generalisation of vocational technology education, Frank Banks and Vanwyk Chikasanda use Bangladesh and Malawi in Chap. 12, ***Technology education and developing countries***, to consider the place of technology education in countries with struggling economies, overpopulation, high unemployment, and low levels of literacy across the population. Poor school attendance is a common problem in such economies, largely due to poverty, inappropriate assessment, teacher-centred pedagogy and an irrelevant curriculum. Education competes with other priorities for scarce resources and educational reforms are decided centrally but implementation is problematic. However, economic reforms emphasising shifts from a rural economy to one that focuses on manufacturing and science and technology-led development raise the profile of technology education, at least politically.

While the rhetoric of *Technology Education for All* in the Global North has been to distinguish it from vocational education, in Bangladesh, Malawi and other emergent economies, the relevance of education to everyday life is paramount and a vocational emphasis might mean that a greater proportion of the population attend school. By providing a curriculum that is relevant, technology education can help reduce poor attendance and high drop out rates, as demonstrated in Chap. 12 by the examples of schools run by the Bangladesh Rural Advancement Committee (BRAC) and the Underprivileged Children Education Programme (UCEP). This chapter therefore offers an important contribution to the book by highlighting the varied challenges faced by technology education, and the dominating influence of the socio-political context on educational opportunities and outcomes.

In Chap. 13, ***Politics and policy***, Kendall N. Starkweather examines the socio-political context of technology education from his background as a technology education professional in the United States, followed by over three decades of experience as an executive with an international technology education association (ITEEA). In order to increase the valuing of technology education by schools and the wider community, including politicians, Kendall reminds us of the need for positive branding. The public needs to be educated about what technology education is, and technology educators must help people trust and believe in the worth of technology education, creating the desire and need for the subject in schools. Alliances with business and industry is often a major disconnect; companies who could benefit from having employees with a technology education background are often unaware of this type of school education. The field of engineering is often complicit in this, requiring students to take mathematics and science courses in order to become an engineer. While these subjects are also important, technology education may be the main reason that a student chooses to become an engineer or technologist in the first place.

Technology education associations have a key role to play in positioning the subject and informing the politics and policy advancing the subject. Teachers who become active benefit from being involved in strategy discussions and resource development, networking advantages, and political support as a result of being known and interacting with others in the technology education and wider education communities. However, as Kendall points out, associations will need to continue to evolve if they are to reflect contemporary values, beliefs and assumptions of the profession, and have robust mechanisms for supporting members to work together in an electronic world.

Anchoring the book in Chap. 14, ***Research challenges for the future***, Marc J. de Vries reviews past research in technology education in order to propose research that may be effective in the context of a developing school subject with uncertain status and a problematic image. He notes that the research in technology education that has developed in a relatively short period of time is remarkable. However, one of the urgent challenges for future research in technology education is to create a better connection with teaching practice. An important possibility here is involving teachers more closely in the research. Second, targeted policy-oriented research is needed and policy makers need to be recognised as an important audience for future technology education research. Third, more sophisticated research is needed on how to better support students' technology learning. To achieve this, Marc argues that a design-based methodology may be particularly fruitful.

Marc concludes his chapter by highlighting the importance of on-going investment in technology education research. Unfortunately, there is a vicious circle here. Technology education research will only be valued when technology education itself is valued—but to provide justification for the appreciation of technology education, relevant research outputs are needed. Indeed, Marc cautions that the future of technology education may depend on the efforts of a few countries that perform extremely well in technology education research and thus provide the ammunition for others to defend both technology education itself and its supporting research to their governments and policy makers. In other words, the extent to which researchers are able to realise closer links between their work and educational practice, and enhance their understanding of policy processes, will likely significantly impact the future of technology education.

It is our hope that this book contributes to the debate by opening up new areas for research and development, and inspiring all who are involved in technology education to continue to work towards curriculum, pedagogy and assessment goals that value students' cultural capital and support their technological learning so that they can contribute meaningfully as current and future citizens.

# Chapter 2
# 'Seeing' and 'Interpreting' the Human-Technology Phenomenon

**Steve Keirl**

This chapter formulates and addresses the dual challenge of how to understand Technology, and how this might inform the development of Technology Education for the future. Philosophy, in its role as critical toolbox, offers hermeneutics as a device for technological interpretation and meaning-making. By setting up a series of binaries, a hermeneutic approach can be used not only to explore Technology holistically and analytically, but also to look to cultural, historical and political relationships. The idea of binarial hermeneutics enables Technology's complexity to become manageable through a focus on issues, which include: local and global, traditional and emergent, product and process, technical and designerly, and academic and practical. These binaries are not 'either-or' but 'at-once-both', enabling analysis along a spectrum. Such a binarial approach offers a way of interrogating the rich and complex areas of Technology Education in order to develop informed ways forward.

## Introduction

This chapter explores the interplay of philosophy, Technology and education and how they relate to curriculum futures. Technology educators have jobs to get on with and curricula to deliver but, as any study of the history of the field will show, ours is not a static field of education. If Technology is our bag then change is inescapably ours too. Understanding change makes us better able to accommodate the future—whether we do so actively or passively! A philosophical engagement with one's field is a powerful way to maintain perspective, to entertain deeper questions and uncertainty, and to think about futures.

S. Keirl (✉)
Design Education at Goldsmiths, University of London, London, UK
e-mail: s.keirl@gold.ac.uk

 
PJ. Williams et al. (eds.), *The Future of Technology Education*, Contemporary Issues in Technology Education, DOI 10.1007/978-981-287-170-1_2

In what follows, *Technology* (big T) addresses the field of Technology-as-phenomenon while *technology/ies* (little t) refers to single or multiple potentially identifiable and specific technologies. *Technology Education* is used as a generic term representing varying formulations and namings for a field of educational activity—which can vary significantly within and across jurisdictions. These varied namings are themselves problematic and cannot be taken to be synonymous.

In this chapter a twofold problem is envisaged—one aspect is about how to explain and understand the rich concept of Technology—how to 'see' Technology—while the other is about how what is seen might inform Technology Education for the future. I think of these problems as *the phenomenon* and *the irony*. Both offer serious challenges to which there are no simple answers. Both signal a need for deeper debate, research and reflection before new settlements can present themselves. In an attempt to address these challenges I introduce an 'interpretive device' to help locate some of the big issues at play.

The *phenomenon* can be put as follows: we humans cannot 'be' without Technology and Technology 'is' by human intention and (inter)action. That is, technologies and humans co-exist intimately. However, despite this intimate human-Technology phenomenon, there is a huge *irony*: why, then, do we not have a parallel education to help understand the phenomenon? Of course we do have Technology Education and its variants around the world and we might argue that it is not their place to wrestle with the complexities of the phenomenon itself but, rather, they should/can only attend to local matters in a timely and manageable way. However, on closer examination, we can see that the phenomenon and practice in schools are not readily separated—or if they are, only a partial Technology Education may be taking place (and I use the term 'partial' here in two senses: of being limited and being biased). We soon see that the phenomenon and the irony both beg special attention through curriculum and teaching.

The challenge is to investigate possibilities that not only help develop our understanding of the phenomenon but also help address the irony. Such investigations can inform future directions for Technology Education. To these ends, the chapter is presented in three broad sections. First, an overview is given of some of the ways in which Technology has been explained by philosophers and theorists who grapple with its complexity. Following the philosophical overview, the interpretive device of *binarial hermeneutics* is introduced. This has been developed with the dual aims of (a) helping identify and locate some key aspects of the human-technology phenomenon and (b) using these aspects to inform 'future directions' that the profession might envisage.

Technology Education from the early years to university, while gradually forming its own philosophical base, still has much to learn from Philosophy of Technology. So, informed by the outcomes of the interpretive device the final section of the chapter looks to some of the educational implications and issues—such as curriculum formulation, teachers' personal professional philosophies, and pedagogical practices—that could be considered on such journeys.

## Philosophy and Theories of Technology—A Brief Overview

> We speakers of English. . .seem to be able to tolerate a high level of ambiguity with respect to our use of the term 'technology'. (Hickman 2001, p. 11)
>
> The philosophy of technology is more like a mosaic of many different ideas and suggestions. Yet, there is a lot that one can learn from this mosaic. Mosaics anyway do have their charm. (de Vries 2005, p. 7)

Despite the prevalence and ubiquity of technologies there is a sense in which they remain largely invisible to us. Once technologies become accepted and cease to be 'new' they are 'normal' and unremarkable and we become uncritical of them. They embed in and form part of our lives, cultures, societies, and our 'being'. So food, clothes, and travel just become part of our way of life rather than being seen as 'technologies'. Technologies cease to be familiar and fade into the background. Similarly, we 'take for granted' the technologies that we are born to—we knew no different so we have no reason to question them. Not only are they part of what we constitute as 'normal' (were we ever to consider them) but they also become our benchmark for what counts as 'old' (before *our* time) and 'new' (not yet experienced).

Technologies shape our lives, our identities, our environments, our cultures, our thinking, our working, indeed our very way/s of being in the world. To what extent is the converse true? Of all the technologies we experience, which ones have we had some say in their introduction, in their design, in their use, or in their positioning of us? Do we really have any control over our technologised being or do we just rather passively adjust our behaviours to accommodate technologies? If we wish to engage such questions, that is, to actively critique technologies, then we are called to explore them from multiple perspectives, interests and uses and in doing so we enter into a world of big questions and issues. Here, the case for education arises.

When it comes to scholarly interrogation, the rich and exciting field of Technology offers a constant dynamic of contestation, possibilities and pitfalls. In some ways, the field's late start in the 'philosophy of' club is due to it having been an object of intellectual curiosity for other academic fields (e.g., history, sociology, psychology, anthropology) thereby having been colonised by them. The field of Technology continues to live somewhat in the shadow of science. In such scenarios the true identity of Technology remains obscured. Furthermore, there are wide differences of understanding both within the Academy and within public discourse about the nature of Technology and how it should be engaged.

Recently, Feenberg (2010), introducing his philosophy of technology said:

> Though we may be competent at using many technologies, most of what we think we know about technology in general is false. Our error stems from the everyday conception of things as separate from each other and from us. In reality technologies belong to an interconnected network the nodes of which cannot exist independently *qua* technologies. . .. It turns out that most of our common sense ideas about technology are wrong. (p. 3)

It is no longer possible to describe technologies simply as 'things' or as 'hi-tech' or as 'applied science' or as 'tools' or as only that which is 'new' or as their being 'neutral'.

In respect of this and to the field's benefit there is a rich and growing number of astute commentators on whom we can draw: Ellul (1964); Feenberg (1991, 1999, 2010); Haraway (1991); Heidegger (1954/1977); Ihde (1979, 1990, 2002); Latour (2007); Lyotard (1979/1984); Mitcham (1994) and Winner (1977, 1986) are a few. Of these, Mitcham offers a comprehensive introductory analysis of Technology's philosophical history, manifestations and interpretations. Elsewhere, Kaplan (2009b) and Scharff and Dusek (2003) offer comprehensive edited collections of readings. All these authors have wrestled with Technology's ubiquity and complexity and have generated invaluable critiques, yet they are far from agreeing on interpretations.

To explore Technology is to engage with multiple philosophical fields: moral philosophy (on ethics and what is 'right'); metaphysics, which includes ontology (on being and existence) and epistemology (on knowledge and knowing); phenomenology (also in some ways on knowledge and on being but especially on subjective experience and consciousness); axiology (on values and aesthetics); political philosophy (on democracy, rights, ideology); philosophy of mind (on intentionality, determinism, mind-body, mind-machine); and, philosophy of language (in a small way used in this chapter, on hermeneutics). However, this is not to suggest that we have to start from such philosophical fields 'cold'. The lived realities of how Technology presents itself in the world offer us multiple alternative starting points.

Critiques of technologies and our relationships with them often present themselves as *issues*, for example: the environment; surveillance; waste; obsolescence; communications; production techniques; genetic engineering; xenotransplantation; identity; democracy; inter-species and environmental justice; consumerism; mechanisation; un/employment; urbanisation; robotics; transport; privacy; and so on—all of which are problematic. There is extensive literature and research on all of these, and more, technological issues.

In the light of such realities it is unsurprising that different approaches to theorising Technology and technologies have emerged that are contributing to debates and bringing new ways of engaging with the field. Four examples, *very* simply presented, serve to illustrate this. First, *critical theory* (e.g., Habermas 1971) with its roots in the Frankfurt School of the 1920s suggests that, as well as looking to identify *what* knowledge should count to make up this or that subject (e.g., technology), we should also look to whose interests are served by the knowledge. Three 'knowledge-constitutive' interests are nominated: the technical (which enables us *to do*); the practical-hermeneutic (which enables us *to understand*); and the critical-emancipatory (which enables us *to be*, that is, to operate as autonomous individuals). These are all readily presenced in and through all technologies: Do we simply use a technology? Do we understand the technology for its social context and the values it embodies? Do we knowingly select, reject and critique technologies? This is not a hierarchy—all three 'interests' matter in education, as can be witnessed in the formulation of technological literacy developed for children in the South Australian Design and Technology curriculum (Department of Education, Training and Employment 2001), that is, it is a Technology Education intended to go beyond the merely technical.

*Actor-network theory* (ANT) (Latour 2007) developed out of studies of complex technological systems. Strictly speaking this is less a theory than a method but the key aspects are that, within any system, human and nonhuman components are attributed equal respect in terms of their significance. Two forms of relations are explored within systems—the material (things) and the semiotic (concepts)—both being key to understanding technologies. ANT, although sometimes charged with vagueness and circularity, certainly offers new insights into the study of human-Technology-social relations, such as understanding how multiple values are attributed to, and by, technologies; and in turn how power is distributed and attributed in technological systems.

For a third example of robust theorising of Technology we can look to Ihde's (1979) phenomenological approaches to the philosophy of Technology advocating what he terms *variational theory* (Ihde 2009; Ihde and Selinger 2003; Sobchack 2006): "…a series of multiple perspectives to recognise the shape, structure, and complexity of the phenomenon (being investigated)" (Eason et al. 2003, p. 125). Such an approach facilitates a richer critique of, and engagement with, Technology and technologies because it recognises notions such as complexity, holism and dynamics (of the phenomenon) alongside the subjective nature of individual and collective perceptions and consciousness. Hermeneutics, simply put as the business of interpretation, is key to phenomenological work.

A fourth theoretical approach is that of *narrative theory,* which Kaplan (2009a) uses to point to the 'narrative' or the 'story' of technologies as a way of 'reading' them. He argues that such reading shouldn't just give a *contextualised* (scene-setting) portrayal of a technology, calling for the *critical* to be ever-present: "[A] critical reading of technology evaluates technical things and systems in terms of their role in achieving social justice and happiness. Technology should…not only be contextualised but read in relation to universalist concepts, such as truth, impartiality, and equality" and he talks of "…narrating things differently to create new ways of seeing the world so that we might imagine, argue for, and create new ways of being in the world" (p. 96). Pointing to simple oppositions that are often posed, he argues that:

> Our choice is not between bad/abstract or good/concrete interpretations of technology, but between conventional readings that leave everything as it is or critical readings that challenge unjust social practices and institutions. Critical narratives connect and relate just as much as they disconnect and interrupt our ordinary contexts of action. They invite us to step back, reflect and deliberate with each other about what is true, right and appropriate and, in so doing, establish the terms of social cooperation. Above all, stories of technology create a common world of meaning for the specialists with technical expertise and the rest of us who just like a good story. (pp. 96–97)

All four of these theoretical approaches offer ways of 'seeing' technologies, and their methods collectively embrace critique, translation, interpretation, reading, describing, explaining, and so on. Like the fields of philosophy, they can help us grasp the richness of the human-technology phenomenon but neither the philosophical nor the theoretical approaches can give us concrete answers or 'truths'. What is worth remembering is that 'seeing' can happen at different levels. For example, we can see rather passively at a surface level but we can also see in deeper ways that expose meaning and bring understanding.

Mitcham's (1994) comprehensive exploration of philosophy of technology issues juxtaposes *engineering philosophy of technology* with *humanities philosophy of technology* and he presents a dialogue between the two. In the spirit of theories, stories and dialogue I'll now introduce some tools to engage with the technological issues. This is done with a technology curriculum for the future firmly in mind.

## Tools to Help Us Understand Technology

What tools might help us not only *see* Technology but also to *better understand* the phenomenon? Might it be possible to develop a method of enquiry that is accessible to researchers and practitioners alike? From what has been said so far, one starting point is to use philosophy itself as a tool. As Hickman (2001) says: "...philosophy is one of the most effective tools we have for tuning up technology" (p. 41). A second tool, coming from within philosophy's toolbox, is that of hermeneutics (the theory of interpretation and understanding). A third device—that of binaries—is used to access Technology's complexity and locate the hermeneutical investigations.

### *Philosophy as a Tool*

Recent economic-rationalist times have seen philosophy being devalued and driven from the Academy. Ironically, the case for its inclusion in education and in everyday discourses may never have been greater. A century ago, Bertrand Russell (1912/1959) had this to say:

> The value of philosophy is...to be sought largely in its very uncertainty. The man [sic] who has no tincture of philosophy goes through life imprisoned in the prejudices derived from common sense, from the habitual beliefs of his age or his nation, and from the convictions which have grown up in his mind without the cooperation or consent of his deliberate reason. To such a man the world tends to become definite, finite, obvious; common objects rouse no questions, and unfamiliar possibilities are contemptuously rejected. (p. 91)

Such statements can be (re-)read with Technology in mind. This is exemplified when, more recently, Quinton (1995) offers this perspective:

> Wherever there is a large idea whose meaning is in some way indeterminate or controversial, so that large statements in which it occurs are hard to support or undermine and stand in unclear logical relations to other beliefs we are comparatively clear about, there is opportunity and point for philosophical reflection. (p. 670)

Otherwise, we can think of philosophy as: the criticism of assumptions; thinking about thinking; critique; analysis or framing; or as an aid to the formation of beliefs, knowledge and ethics. Equally, philosophy offers an excellent playground for the development of new theories or new ways of looking at phenomena. This is not to suggest that theorising is not highly practical—it can be a means to an end. As Singer (1993) says of ethics:

> Ethics is practical, or it is not really ethical. If it is no good in practice, it is no good in theory either. Getting rid of the idea that an ethical life must consist of absolute obedience to some short and simple set of moral rules makes it easier to avoid the trap of an unworkable ethic. (p. 204)

## *Hermeneutics as a Tool*

Historically, hermeneutics was concerned with the interpretation of religious texts to establish what meaning they carried and what the whole–parts relationships of the text might be. However, over the past century hermeneutics has moved well beyond text as its subject matter and has also refined and deepened its methodological approaches (see, for example, Bohman 1999; Gadamer 1977, 1975/2004; Habermas 1971; Mitcham 1994; Palmer 1969).

To work hermeneutically is not only to explore holistically and analytically, but also to look to cultural, historical and political relationships. Having grown from more formal, language-based methods of interpretation, hermeneutics has become an actual existential event for the interpreter. That is, hermeneutics is today seen as much for how the hermeneutic act itself shapes us as for how it serves as an interpretive tool.

When we work hermeneutically, understanding comes of one's own historical and cultural positioning and new possibilities present themselves to us. In hermeneutics, all of analysis, synthesis, critique, judgement, dialectical and logical reasoning, and reflective practice (with oneself and with others) combine to bring new understandings. The familiar is made strange and new ways of seeing emerge. In all these ways, hermeneutics is about *both* interpretation *and* understanding.

We are at once *a part of* the hermeneutic act (subjectively) and *not apart from it* in an 'objective' sense. There is a sense that we are in a reflective conversation with the phenomenon (in our case, Technology). Thus, we look to how the phenomenon 'speaks' to us but we *reflect back to* and *reflect on* what we perceive. The 'conversation' that we have will almost certainly not be the 'conversation' that others have with the same phenomenon. In this way, when we understand that technologies are contradictory, arbitrary, controversial or contested it is because our hermeneutic (and ontological) experiences differ from those of others. Multiple reciprocations take place. Through hermeneutic engagements we not only get to understand the phenomenon of Technology better but we also get to understand ourselves (personally and collectively) better too.

When Mitcham (1994) counterposes his engineering philosophy of Technology with his humanities philosophy of Technology he reminds us of the significance of hermeneutics: "...because of the central place interpretation occupies in all such humanities reflection...The hermeneutic or interpretive enterprise is pervaded by personal, interpersonal, and historically conditioned elements, and thus tenuously articulated within a human world of fluctuating intersubjective consensus" (p. 63). Hermeneutic work around technologies can take us away from the mythologies that come with the field (e.g., technologies as 'neutral', 'hi-tech', 'applied science', 'new', 'good', etc.) and it becomes possible to develop new language, terminologies, theories and analyses.

Given the complexity of Technology and the vast multiplicity of technologies, where could appropriate hermeneutic investigations begin? How could they be located? A clue comes from Gadamer (1975/2004), who reminds us that "Hermeneutic work is based on a polarity of familiarity and strangeness" and that "*(t)he true locus of hermeneutics is (the) in-between*" (p. 295, original emphasis). This brings us to the use of binaries.

## *Using Binaries as a Device to Locate Hermeneutic Work*

First, some clarification of terms... The question could be asked: "Why binaries, why not dualisms?" The answer lies in the meanings and varied applications of the respective terms. Whilst 'binary' in mathematics means 'having a base of two' it has also acquired a popular sense of *either-or*, which is in fact what dualism means. Dualism occurs in philosophical discourse with particular reference to two *distinct* things—the mind-matter dualism of Descartes being the classic example. In contrast, binary means *both-at-once, two-together*, a *compound* or, perhaps, a *co-dependence* (which all serve well the intentions of this chapter in eschewing polarisation and/or separation of entities, phenomena or positions). Where dualism is about distinction, binary is about indistinction. This validates the hermeneutic approach.

It is also helpful to say something about dialectics and hermeneutics. In the Kantian distinction between analytic logic and dialectic logic, the latter is less about facts than about a style of reasoning. For Hegel, dialectics was a method of overcoming (seeming) opposites—sometimes expressed in simple form as: [a] thesis posed, [b] antithesis counterposed, and [c] synthesis (drawn out of the arguments of both [a] and [b])—creating a new settlement. While dialectics is a most appropriate tool for addressing incompatible dualisms, dialectic methods can be fruitful in hermeneutic enquiry. However, as has been indicated, dialectic reasoning is not the only hermeneutic method.

If the phenomenon of Technology is complex and the number of technologies is immeasurable then a way is needed to try to address, frame or manage the problem. This is *not* about reducing the problem. While hermeneutics offers engagements with complexity, using binaries is a means of focussing the approach. The idea is to use binaries as sites to locate, expose or invite hermeneutic enquiry. Using binaries allows us to capture or signal a range of issues that we may wish to address (including some that might appear in Technology Education). The nomination of any binary intentionally *foregrounds one aspect of Technology* while backgrounding (but still accommodating) others. In short, the binaries locate spectra of issues while the hermeneutics bring forward interpretations across the spectra.

To give an example... (All the binaries that follow are expressed as 'Technology as *at-once-both...*' to remind us of the co-dependence issue.)

We can set up a binary of 'Technology as *at-once-both arts and science*'. If we try to say that technology is *only* arts (perhaps in the sense of crafting and creativity) or that it is *only* science (perhaps in the sense of objective study) we come unstuck because we cannot argue the exclusivity of one over the other. On the hermeneutic journey we might explore: what constitutes a science or an art; in what ways technology reveals itself to us as art, as science; whether technology is 'applied science', a branch of science, or (after Lueckenhausen 1989) is art made useful; Mitcham's (1994) juxtaposition of engineering with humanities; whether/how art and science meet in technology; how a technology can be both science and art at once; and so on.

What the hermeneutic journey opens up here are questions of: epistemology, where technique sits, where Design fits, the nature of creativity, how we understand disciplines, form and function (or aesthetics and instrumentalism), human experience, objectivity-subjectivity, and more. By engaging with the binarial issues present across the arts-science spectrum we work hermeneutically and discover new ways of seeing and understanding that which we think we 'know'. The outcome would be deeper understandings of the nuances of Technology, as well as clearer understandings of whatever we mean by 'arts' and 'science' *and* whether such a binary is helpful in exploring Technology. The educational point is not to resolve a dualism but to learn from the understandings and meanings that develop from the hermeneutical practice—to interpret fruitfully.

Subsequently, understandings gained from the hermeneutic explorations of any big-T Technology binary can also be *tested and refined* by applying case studies of particular small-t technologies to triangulate findings, refine thinking, and imagine new possibilities. For example, the submission of a washing-up brush, an aeroplane or a bridge to hermeneutic scrutiny within the arts-science binary can validate the enquiry as well as help clarify philosophical thinking.

## Technology's Binaries—Putting Binarial Hermeneutics to Work

When the three tools of philosophy, hermeneutics and binaries are gathered to frame and investigate the philosophical dimensions of the phenomenon of Technology, the term given for the practice is *binarial hermeneutics*. Selected binaries set the context, and a hermeneutic approach invites interpretations and revised understandings of philosophical interplay with technologies. By investigating a range of binaries it becomes possible to identify technological-philosophical threads or themes for further critique and analysis.

The binaries that follow are signifiers of Technology discourses in which there are multiple possible positions—they echo the *arbitrary* nature of the phenomenon of Technology and the notion of technologies being *multistable* (Ihde 2002; Ihde and Selinger 2003) or *polypotent* (Sclove 1995). There is nothing sacrosanct about the binaries—they are starting points and there are certainly other binaries which could be nominated. Further, the binaries are not necessarily qualitatively the same—some allude as much to informed (or ill-informed) public discourse as they do to orthodox philosophical enquiry.

## The Arch-Binary of At-Once-Both Human and Technology

It occurs that there is one *arch-binary* that stands out as the epitome of the challenges under investigation. This was expressed in the chapter's introduction and it is the binary of *at-once-both human and technology*. Simply put, the phenomenon of *Human* cannot *be* without reference to technologies and the phenomenon of *Technology* cannot *be* without reference to humans. This binary is the arch-binary because it is the arena for the acting out of all other binaries. As such, it is also the arena of most of the major philosophical discourses—epistemology, ethics, metaphysics (including identity, ontology and religion), philosophy of mind (including states of mind, mind-body, mind-machine, the will, determinism, and intentionality), and political philosophy.

There are many entry points to this arch-binary. Some possible explorations might be:

- simply expressed, through the question: how human are technologies, how technological are humans?
- around identity, free will and the ways in which technologies and humans shape each other;
- analysing decision making: whether human choice-making, human design decisions, or technologically programmed decision-making (e.g., genetic, robotic, digital);
- examining *transhuman* and *posthuman* (in the technologically-framed sense) scenarios that, tenable or not, are both culturally and politically significant while also offering unknown futures (Bostrom 2009; Broderick 2001; Kurzweil 2005);
- reflecting on Foucault's (1989/2000) postmodern reminder that 'man' is 'a recent invention';
- considering whether *humanity*, *human-beingness*, and *humanism* are constructs that may not be sustainable (posthumanity in the postmodern framing, see Badmington 2000);
- critiquing Kurzweil's (2005) view that 'technology is evolution by other means' and understanding scenarios of soft and hard posthumanism (technological, postmodern or otherwise).

Despite the postmodern terminology, what is presented here indicates the kinds of debates and issues that are needed to accompany our technologically sophisticated developments (an insipid phrase in the circumstances, perhaps). As emergent technologies combine and interact in new ways and with humans (and other species and environments), so multiple aspects of human practice change and new discourses are warranted. In line with technological shifts there are strong grounds for parallel shifts in educational, professional, political and social discourses—our future shared directions.

Such discourses could conceptualise *Technological-being* on a level footing with *Human-being*—to treat both phenomena with some 'symmetry' (drawing on ANT and the constructivist principle of symmetry as it can apply to the study of technological controversies). In such a way the human and the technological might each be scrutinised and celebrated for what it gives to and takes from the other. This would be a matter of continuous reciprocal critique rather than an attempt at simple balancing of the books of bias or of power distribution. Ultimately, hermeneutic imaginings of the *degree* of qualitative, as well as quantitative, merger of the two is a scenario in which education, formal and otherwise, could play a key role.

What, then, might other candidate binaries be? First, a range of binaries of Technology is presented.

## At-Once-Both Visible and Invisible

This has been signalled earlier. When technologies become so accepted, so unquestioned that they become almost invisible (that they are everywhere yet nowhere at once) does a taken-for-grantedness occur? To what extent is the phenomenon questioned, or do we passively accept our technological circumstances? What are the disruptions to such circumstances that remind us of what has become invisible—major catastrophes, shortages, climate issues, disruptive technologies? Is the invisibility of the everyday matched by an invisibility of our evolution? How might the foreground-background question (in phenomenology) be considered? In other words, when are our technologies present with us, foregrounded and part of our active existence/being? When are they backgrounded and part of our passive existence/being? What do we 'see' as technology—are education, law, language, 'the arts' all technologies? Philosophical questions around epistemology, existence and identity are engaged here.

## At-Once-Both Positivist and Antipositivist

The *seemingly* tangible nature of technologies and traditional ways of assessing them ('Does it work/does it do the job?') have meant that Technology has found itself framed as addressing the instrumental and the material and being aligned with science. This has been countered by antipositivism, which resists what is seen as simplification and narrowness of interpretation. However, in its most radical form, antipositivism has been charged with creating mires of relativism that are difficult to penetrate. While technology is just 'obvious' to some, to others it is nebulous.

Between any positivist and antipositivist extremes lie multiple epistemological issues and interests. Hermeneutic work would unveil varying 'knowledge constructions' of both simple and contested types—constructivist, propositional/procedural, situated, tacit, experiential, embedded, dis/interested, etc.—collectively, in their postmodern articulation, 'knowledges'. Questions of what is valid knowledge and what constitutes valuable knowledge present themselves—as do engagements with reductionism and holism.

## At-Once-Both Utopian and Dystopian

Technology is basically good. Technology is basically bad. Here philosophical questions arise around values, existence, ethics, post/humanism, determinism, and eco-philosophy. Does Technology result from the realisation of our best (design) intentions or are the intentions not actually 'good'? Why are there unintended consequences from technological developments? How might hermeneutic enquiry explore questions of control, free will, power distribution, psychological wellbeing, discourses around wellness, optimism, and pessimism? What is the place of critiques of markets, production, waste, sustainability and surveillance? What is the role of technological rejection, cynicism, scepticism, or faith? What is the Technology-Nature relationship? Is there dissonance between human progress and technological progress? What of Frankenstein and Faust? Do technologies enable us and enrich us or are we 'encapsulated in artifice' (Nye 2007)?

## At-Once-Both Subject and Object

Technology provides us with rich experiences—quantitatively and qualitatively—but can we clarify what we know (or think we know) about technological objects and experiences and, further, how what others know and experience is the same as us? Humans and technologies are capable of being either subject or object or (through the binary) always-both. In ANT, "Objects too have agency" (Latour 2007, p. 63). Thus, within systems, who/what is the subject/object? For emerging technologies, what do we really know about how they will shape/interact with ourselves, others, other objects and systems? To what uses can we put a phenomenological hermeneutics (of perceptions and experiences) of human-technology relations (Ihde 1990)?

## *At-Once-Both Democratic and Non-democratic*

How do technologies serve or deny democracy? What are the citizenship issues? Who has access to data? Does our use of Technology lessen the quality of life of others? At what point in a technology's development is ethical critique or democratic engagement allowed, for example, at pre-conception, at the design phase, during creation, after manifestation? (Keirl 2009). Philosophical questions arise around politics, ethics, existentialism, determinism (and volition). Who is empowered, disempowered? What is the role of education? Should decision-making and guidance be left to experts or a technocracy? Are our engagements with technologies open and free or does Technology entrap and constrain?

## *At-Once-Both Generic and Specific*

Here the Technology/technologies interplay begs interrogation to open up epistemological, ethical and values issues. To what extent can the specific inform the general or the general the specific? Can there be any technological rules or propositions that might help us understand the phenomenon? Quantitatively or qualitatively are there particular or general values positions that can be taken regarding all technologies? Can we 'know' certain things about all technologies or only some things about some technologies? Hermeneutically, are there ways that technologies 'speak to us' either generally or specifically?

## *At-Once-Both Modern and Postmodern*

While Technology is often framed as 'modern', postmodernism does much to challenge many of the 'givens' of Technology: the idea of 'progress'; of technological determinism; that there is one form of technological knowledge rather than multiple knowledges; that rationalism and optimism guarantee outcomes; and, further, it presents the case that there are no 'grand narratives' (Lyotard 1979/1984) or universal rules so far as Technology is concerned. While modernist faith in perceived technological certainties can be espoused, postmodernism might be celebrated for its ways of creating mystery, being reactive and critical, being playful and celebrating the absurd about technologies (Kellner 1991; Lather 1991).

## *At-Once-Both Natural and Artificial*

While it seems 'natural' for us to be creative and to act technologically upon the world there are clearly ways that such actions work against nature. Is it just a matter of genetic disposition to 'use' nature for our purposes? Once we have created a technology (whether or not we draw upon natural resources) is the creation a solely artificial entity? What are our relationships with other species and environments? Is the only way we think and behave with regard to Technology anthropocentric? Taking Franklin's (2004) lead, how should we consider the biosphere-bitsphere relationship? What understandings can be gained from how cultures and faiths other than our own interact with technologies (Diamond 1998; Kraybill 1989/2001)? Are we creating, through our technological practices, behaviours and cultures that are further distancing us from nature? In working hermeneutically in this binary we engage with values analysis, epistemology, existentialism, phenomenology, and eco-philosophy.

Having presented a flavour of the potential of binarial hermeneutics for helping us understand Technology, how can such an approach be brought to Design and Technology education?

## Technology Education Binaries

The idea of binarial hermeneutics can work well to inform the curriculum—whether through curriculum policy-making or through the delivered curriculum. In some ways the educational binaries reflect those for *Technology* and *technologies*. However, there are two senses in which the binaries for the educational context gain some piquancy. First, their resolution is now towards curriculum action rather than one of philosophical-hermeneutic reflection. However, we should note that the better our philosophical-hermeneutic explorations in the Technology-technologies arena, the better equipped we are to resolve the educational challenges. Second, matters become more explicitly political where curriculum arbitration is concerned since the binaries address questions of what experiences and learning students should gain.

We should also note that some of the binaries are applicable across the curriculum, that is, beyond Technology Education. They are included here precisely because of the particular way that the binary applies to Technology Education. This becomes an important part of building the integrity of Design and Technology in educational circles—it is a field of unique challenges and special circumstances that cannot be dealt with through blanket policy-making. These points made, what are the Technology Education binaries that present themselves for hermeneutical enquiry? The following, as with the Technology binaries, is a selection, they interplay, and other possibilities could be nominated.

### *At-Once-Both Status Quo and Change Agent*

This is a classic binary for education itself. Is the role of the school to maintain the status quo or to bring about change? Technology Education's particular challenges are strongly evident here with shifting social and workplace practices and new technologies constantly evolving. Curriculum, schools and the profession find themselves continuously facing change and adaptation. Which techniques and technologies are to be valued or abandoned? Should Technology Education be taught uncritically or critically? Should social issues and challenges be engaged or bypassed?

### *At-Once-Both Local and Global in Perspective*

Is the curriculum inward- or outward-looking? Does a local curriculum operate, that is, is it determined locally around the community's needs? Or is curriculum determination centralised and controlled? Whatever the organisational control, what international and global perspectives are articulated in the curriculum? In taking such perspectives, what kinds of contradictions around resources, values and ethics arise?

### *At-Once-Both Traditional and Emergent Technologies*

Is the curriculum crafts-based, existentially sedentary, low production and low-tech—or hi-tech, existence-changing and (r)evolutionary (including technologies not yet disseminated in the community)? Whether new or old, do technologies serve the mass or only elites? What spectrum of technologies is celebrated and studied to show how lifestyles are shaped? To what extent are 'old' or 'new' technologies de/valued? What comparisons can be made between established and emergent technologies on the basis of costs, uses and consequences?

### *At-Once-Both Product and Process (For the Teacher)*

With regard to the *pedagogy* of Technology Education, is learning best addressed through the creation of products—emphasising, say, technique, efficiency, quality, production methods and standards? Or is it a matter of process where students learn designerly behaviours and dispositions to work creatively in teams and alone? Is the 'output' of education to be the capacity to (re)produce or the capacity to adapt and (re)imagine new possibilities?

## At-Once-Both Technical and Designerly (For the Student)

Is Technology Education about technical skilling alone—that is, in simply learning how to use tools or software? Or is it about a more embracing curriculum of critical-designerly behaviours for being-in-the-world—those that would serve the hermeneutical dispositions of the students? Should such education be restricted to creativity and imagination rather than technical can-do? Thus, is education to focus on a limited repertoire of skills developed in depth or is it to be more diverse and holistic in its practice? What kinds of knowledge are to be valued? Is Technology Education a seeding ground for student self-expression and identity formation?

## At-Once-Both Instrumental and Liberal (For Society)

This binary is well articulated by Layton's (1994) research into the stakeholder interests in Technology Education. Is the primary aim to serve the needs of the economy or is there something special that Technology Education can bring to all students: "...a distinctive form of cognition, unique and irreducible. As such, all children should have access to it, as a matter of right and in order to develop their full human potential" (p. 17)? Are some goals for short-term employability and specific industrial and business needs while others are to create an educated citizenry?

## At-Once-Both for Society and Student

The two binaries above can be considered in another way by juxtaposing the needs of the student with the needs of society at large (indeed, these days, we should perhaps expand the notion of society to embrace global citizenry). Is Technology Education about giving the richest and most fulfilling experience to all students or is it to meet society's needs for economic productivity, critical consumers, and active citizens?

## At-Once-Both Academic and Practical

Although this framing should be a quirk of history in the twenty-first century, in many societies the hands-head divide (with no place for heart) remains embedded in educational culture dating from Ancient Greece and it materialises still in curriculum contestations that would see aspects of technology learning split amongst workshop-based manual arts and crafts or industrial technology, laboratory-based theoretical/experimental science (materials, testing, problem-solving), and classroom-based language-rich technological social issues in social studies. Epistemological (let alone ethical) defences for separation within this binary are weak but the cultural traditions remain strong.

## *At-Once-Both Cross-Curricular and Subject*

Several of the above binaries resonate with this classic curriculum organisation binary. How should Technology Education be formulated to address both Technology and technologies? It would seem that to deliver Technology Education as a *subject* would demand its being positioned at particular extremes of several of the binarial spectra—if only on the grounds of efficiency. Looking beyond the subject-alone position, while the arguments for some kind of broad technological literacy for all students are gaining strength, awareness is needed of the spectrum of framings of 'technological literacy' that exist. These can be from the technical to the critical-emancipatory (see, for example, Dakers 2006; Keirl 2006) and could serve widely differing agendas. Whatever the framing, the delivery of any comprehensive technological literacy is unlikely to be possible through a single subject or learning area but, rather, would become the business of every teacher and department in a school.

Hermeneutic understandings gained from this binary reflect some of the major issues facing curriculum designers and facilitators today. When the relationships between the general and the particular have to be considered, they are as apparent in the cross-curriculum/subject question as they are in the special nuances of Technology Education. Further, there are epistemological matters at play here when we have to arbitrate between holistic understandings of knowledge (constructivist, experiential, integrated) used thematically and subject-disciplinary knowledge.

Despite the apparent overlap between some of these binaries, any move to combine or merge them would defeat the whole point of 'opening up' new understandings via hermeneutic processes. There are many nuances at play and they matter. To reiterate, it is erroneous to see any binary as a dualistic 'either-or'—that would be a form of reductionism. Design and Technology's curriculum challenge is the management of Technology as holism while also addressing what counts as appropriate with regard to individual technologies/techniques. The binaries offer an approach to managing the holism of Technology's (and technologies') complex and contested values.

## Conclusions...

A binarial hermeneutic method has been used to locate and explore ways of 'seeing' and 'interpreting' the complexity of the Human-Technology *phenomenon* as well as to offer a way of looking at Technology Education and its curriculum. It is important to remember that Philosophy of Technology is itself nascent and that much (mis)use is still made of culturally embedded myths and misunderstandings to explain what we think 'T/technology' is. Thus emerges the very educational need (the *irony* as I have construed it) that this book seeks to address. What might the future directions be for Technology Education?

While philosophy has been marginalised in the Academy in recent times and has certainly been dropped from many teacher education courses, I have tried to show something of its significance in helping frame the educational challenge. The key use of philosophy is as a tool—as a way of interrogating the status quo and introducing alternatives. However, as the binaries show, multiple values positions compete with each other and not all can be accommodated. If true *education* is the goal, then some tricky weighing up is needed if educational quality is to be maintained. Meanwhile, quantitatively, the crowded curriculum can only allow for so much. Sophisticated judgements are called for to consider both what should constitute 'good' Technology Education and how it should be articulated.

There are no signs that there is (or will be) a stable school of philosophical thought to which Design and Technology educators might relate. It is likely that there will always be multiple schools of thought—a simple reflection of how intimately the human-Technology experience is woven. In such a situation it is a matter for leading thinkers and practitioners in the field of Technology Education to assemble the most comprehensive overview of the issues as well as to consider how these might be best addressed. To such an end, the following are offered for consideration:

- there is an ever-growing body of literature (both from Philosophy of Technology and from Technology Education) which should be drawn upon—for the two fields respectively, Mitcham (1994) and de Vries (2005) are excellent starting points;
- both broad fields of enquiry (Technology and Education) are increasingly engaging discourses of postmodern critique, constructivism, feminist theory, phenomenology, and cultural studies;
- common to both fields is substantial work around critical theory and its application. In Education: enlightened curriculum iterations, curriculum design texts (e.g., Smith and Lovat 1991), the theorising of Freire, and the critical literacy movement. In Technology notably Habermas and Feenberg;
- design is key to any technology's coming into being yet remains somewhat invisible in Technology Education—a concern, given its potency as a vehicle of learning and for pedagogical practice;
- are technology curriculum wars a realistic possibility (Keirl 2012)? Might a division of ideological proportions create a split in the thinking and the organisation of the field? If so, is this a problem or a sign of good educational health?
- is a way forward to identify and agree upon a single robust strategy for the philosophical framing of Technology Education? One example might be to

focus on the three 'E's of *epistemology, ethics*, and *existentialism* articulated through a pedagogy and curriculum based on design and critical theory; and,
- of course, there are others...!

## *On a Political Note...*

Education is a political act and in broad terms can be found in one of three forms across jurisdictions: neo-liberal, liberal/progressive, and socially critical. How Technology Education frames itself is a matter of political philosophy too. How Technology Education responds/reacts to the kinds of options presented through the binarial hermeneutics or, alternatively, which interpretations it chooses to apply will in most aspects amount to being a political act. Commonly, this will be because of education's role in relation to the state but is also simply a function of any educator's decisions about what constitutes education.

## *Curriculum Scenarios*

I shall close with three possible prospective curriculum scenarios for Technology Education (they are by no means exclusive). They are:

- the dissolutionist case—that there are strong grounds to abandon the idea of Technology Education in any overtly identifiable form (e.g., as a subject) and that technological knowledge/learning can/should be adequately addressed elsewhere in the curriculum, for example, across the Arts, in Science, and in Social Studies;
- the specialist case—that there are strong grounds for a subject or learning area called (Design and) Technology Education to play a full part in a future-focussed curriculum and that this subject should be compulsory for all students throughout the compulsory school years;
- the comprehensive case—that there are strong grounds for both a subject called Technology Education and a formal articulation of technological literacy across the curriculum. This arrangement would utilise all of (a) the subject approach; (b) integrated cross-curricular project work with all other subjects/departments; and, (c) have a high level of technological literacy addressed though all other school activities.

While what has been set out in this chapter may seem remote *from* the classroom, it is my view that the issues are ever-present *in* the classroom. The essential case (the *irony*) is that the remarkable phenomenon called Technology is not at all well articulated by a matching Technology Education. Given that significant amounts of philosophical activity generate around the arch-binary of the *at-once-both Human and Technology*, it seems plausible that Technology Education should in turn seek to be richly informed philosophically. A way forward for this has been offered using the triple tools of philosophy, hermeneutics and binaries—in combination. Anyone reading this chapter will hold a personal set of values and, one way or another, will hold values-positions on Technology and on Education. No less is the case for each Design and Technology teacher who, in their very pedagogy and curriculum planning, is an arbiter of the competing values that have been presented here in the binaries.

The Future, our personal and collective futures, and technologies, will all continue to become richer and more complex, and this will remain Education's challenge. This is Design and Technology Education's future and we must engage with it responsibly and in informed ways. Many of the issues are both age-old and perennial and, while technologies change and evolve, the underlying philosophical issues do not. Whatever our role in the future of Technology Education, we have good reason to be positive and optimistic about it and about our capacities to address its considerable challenges. However, to ignore the philosophical dimension and its fundamental significance would not seem advisable.

## References


Badmington, N. (Ed.). (2000). *Posthumanism*. Basingstoke: Palgrave.

Bohman, J. (1999). Hermeneutics. In R. Audi (Ed.), *The Cambridge dictionary of philosophy* (2nd ed., p. 378). Cambridge: Cambridge University.

Bostrom, N. (2009). The future of humanity. In J. K. B. Olsen, E. Selinger, & S. Riis (Eds.), *New waves in philosophy of technology* (pp. 186–215). Basingstoke: Palgrave Macmillan.

Broderick, D. (2001). *The spike: How our lives are being transformed by rapidly advancing technologies*. New York: Forge.

Dakers, J. R. (Ed.). (2006). *Defining technological literacy: Towards an epistemological framework*. Basingstoke: Palgrave Macmillan.

de Vries, M. J. (2005). *Teaching about technology: An introduction to the philosophy of technology for non-philosophers*. Dordrecht: Springer.

Department of Education, Training and Employment. (2001). *South Australian curriculum standards and accountability framework* (SACSA). Retrieved from http://www.sacsa.sa.edu.au

Diamond, J. (1998). *Guns, germs and steel: A short history of everybody for the last 13000 years*. London: Vintage.

Eason, R., Hubbell, J., Jørgenssen, J. F., Mallavarapu, S., Plevris, N., & Selinger, E. (2003). Interview with Don Ihde. In D. Ihde & E. Selinger (Eds.), *Chasing technoscience: Matrix for materiality* (pp. 117–130). Bloomington: Indiana University.

Ellul, J. (1964). *The technological society*. New York: Alfred Knopf.

Feenberg, A. (1991). *Critical theory of technology*. Oxford: Oxford University.

Feenberg, A. (1999). *Questioning technology*. London: Routledge.

Feenberg, A. (2010). *Between reason and experience: Essays in modernity and technology*. Cambridge, MA: MIT.

Foucault, M. (1989/2000). The order of things: An archaeology of the human sciences. In N. Badmington (Ed.), *Posthumanism* (pp. 27–29). Basingstoke: Palgrave.
Franklin, U. M. (2004). *The real world of technology*. Toronto: Anansi.
Gadamer, H.-G. (1975/2004). *Truth and method* (J. Weinsheimer & D. G. Marshall, Trans.). London: Continuum.
Gadamer, H.-G. (1977). *Philosophical hermeneutics* (D. E. Linge, Trans.). Berkley: University of California.
Habermas, J. (1971). *Knowledge and human interests*. Boston: Beacon.
Haraway, D. (1991). *Simians, cyborgs and women: The reinvention of nature*. New York: Routledge.
Heidegger, M. (1954/1977). *The question concerning technology and other essays* (W. Lovitt, Trans.). New York: Harper & Row.
Hickman, L. A. (2001). *Philosophical tools for technological culture: Putting pragmatism to work*. Bloomington: Indiana University.
Ihde, D. (1979). *Technics and praxis*. Dordrecht: Reidel.
Ihde, D. (1990). *Technology and the lifeworld: From garden to earth*. Bloomington: Indiana University.
Ihde, D. (2002). *Bodies in technology*. Minneapolis: University of Minnesota.
Ihde, D. (2009). *Postphenomenology and technoscience: The Peking University lectures*. Albany: State University of New York.
Ihde, D., & Selinger, E. (Eds.). (2003). *Chasing technoscience: Matrix for materiality*. Bloomington: Indiana University.
Kaplan, D. M. (2009a). How to read technology critically. In J. K. B. Olsen, E. Selinger, & S. Riis (Eds.), *New waves in philosophy of technology* (pp. 83–99). Basingstoke: Palgrave Macmillan.
Kaplan, D. M. (Ed.). (2009b). *Readings in the philosophy of technology*. Lanham: Rowman and Littlefield.
Keirl, S. (2006). Ethical technological literacy as democratic curriculum keystone. In J. R. Dakers (Ed.), *Defining technological literacy: Towards an epistemological framework* (pp. 81–102). Basingstoke: Palgrave Macmillan.
Keirl, S. (2009). Seeing technology through five phases: A theoretical framing to articulate holism, ethics and critique in, and for, technological literacy. *Design and Technology Education: An International Journal, 14*(3), 37–46.
Keirl, S. (2012). Technology education as "controversy celebrated" in the cause of democratic education. In T. Ginner, J. Hallström, & M. Hulten (Eds.), *Technology education in the 21st century: Proceedings of the PATT 26 conference* (pp. 239–246). Linköping: Linköping University.
Kellner, D. (1991). Reading images critically: Toward a postmodern pedagogy. In H. A. Giroux (Ed.), *Postmodernism, feminism, and cultural politics: Redrawing educational boundaries* (pp. 60–82). Albany: State University of New York.
Kraybill, D. B. (1989/2001). *The riddle of Amish culture*. Baltimore: John Hopkins University.
Kurzweil, R. (2005). *The singularity is near: When humans transcend biology*. London: Penguin.
Lather, P. (1991). *Getting smart: Feminist research and pedagogy with/in the postmodern*. New York: Routledge.
Latour, B. (2007). *Reassembling the social: An introduction to Actor-Network-Theory*. Oxford: Oxford University.
Layton, D. (Ed.). (1994). *Innovations in science and technology education* (Vol. V). Paris: UNESCO.
Lueckenhausen, H. (1989). Design is art made useful. *Australian Art Education, 13*(2), 33–37.
Lyotard, J.-F. (1979/1984). *The post-modern condition: A report on knowledge*. Manchester: Manchester University.
Mitcham, C. (1994). *Thinking through technology: The path between engineering and philosophy*. Chicago: University of Chicago.
Nye, D. E. (2007). *Technology matters: Questions to live with*. Cambridge, MA: MIT.

Palmer, R. E. (1969). *Hermeneutics: Interpretation Theory in Schleiermacher, Dithey, Heidegger, and Gadamer*. Evanston: Northwestern University.
Quinton, A. (1995). Philosophy. In T. Honderich (Ed.), *The Oxford companion to philosophy* (p. 670). Oxford: Oxford University.
Russell, B. (1912/1959). *The problems of philosophy*. Oxford: Oxford University.
Scharff, R. C., & Dusek, V. (Eds.). (2003). *Philosophy of technology: The technological condition—An anthology*. Oxford: Blackwell.
Sclove, R. E. (1995). *Democracy and technology*. New York: Guilford.
Singer, P. (1993). *Practical ethics*. Cambridge: Cambridge University.
Smith, D. L., & Lovat, T. J. (1991). *Curriculum: Action on reflection*. Sydney: Social Science Press.
Sobchack, V. (2006). Simple grounds: At home with experience. In E. Selinger (Ed.), *Postphenomenology: A critical companion to Ihde* (pp. 13–19). Albany: State University of New York.
Winner, L. (1977). *Autonomous technology: Technics-out-of-control as a theme in political thought*. Cambridge, MA: MIT.
Winner, L. (1986). *The whale and the reactor*. Chicago: University of Chicago.

# Chapter 3
# Theorising Technology Education from a Cultural-Historical Perspective: Foundations and Future Imaginings

**Marilyn Fleer**

In this chapter, three interrelated concepts drawn from cultural-historical theory are interrogated to develop a futures orientation to technology education. They are: tools and signs as cultural practice, everyday concepts and technological concepts in technology education, and imagination and creativity in design and technology education. Bringing these together in technology education offers a way for students and teachers to together consider the way the cultural and historical (and future) ecology of societies is realised through the knowledge base and values-oriented activity of design and technology, and to understand that technological activity is embedded in a deep knowledge of self and community. As such, technology education has both an ethical and moral responsibility to support imaginings that sustain people and communities in harmony and for the well being of the broader ecological and social environment.

## Introduction

The foundational concepts inherent in the practices of technology education augur well for the current international need for societies to have creative and imaginative thinkers who formulate and design possibilities not yet imagined. The significance of this need is huge, but as yet this has not been fully realised by governments or by country specific curriculum boards who place technology low on their priority list. The classical mind-body split still foregrounds the beliefs and practices found in many education contexts, hampering directions towards a futures-focused re-conceptualisation of technology education. To help move the agenda forward, this chapter discusses concepts from cultural-historical theory in order to build

M. Fleer (✉)
Early Childhood Education, Monash University, Melbourne, Australia
e-mail: marilyn.fleer@monash.edu



PJ. Williams et al. (eds.), *The Future of Technology Education*, Contemporary Issues in Technology Education, DOI 10.1007/978-981-287-170-1_3

foundational knowledge of technology education where imagination and creativity are central features for practice and where a futures orientation is theorised.

To achieve the re-conceptualisation of technological practices with a futures orientation, new concepts are needed. The constructs of *tool and sign*, *everyday and technological concepts*, and *imagination* (abstract) *and creativity* (practice) are used in this chapter for analysing the past, present and future practices in technology. These concepts are introduced in the first part of the chapter for building an epistemological basis for technology education. This is followed by a discussion of these concepts in the context of the literature that has evolved since they were first mooted by Vygotsky. In the latter part of the chapter, these concepts are used to actively contribute to re-conceptualising technological practices and constructs with a futures orientation. Through this, foundational knowledge is built and a new theory for a futures orientation to technology education is presented.

A cultural-historical, or sociocultural, conception of technology takes into account a system of concepts that together realise a futures perspective for the field that is transformative of not only the curriculum, but also the students who engage in these concepts and practices. Underpinning cultural-historical theory is dialectical logic. *Dialectical* is understood in this chapter as the "*history of science and technique* collectively created by people" (Ilyenkov 2003, p. 102) where both the past and the future are enacted in the present—as a dialectic. The future does not exist if there is no past; similarly, the past does not exist if we do not consider the future.

In a dialectical conception, past and future are realised "both-at-once" in the present (see Keirl, Chap. 2, this volume).[1] Because dialetical logic actively dispenses with the traditional mind-body divide that has plagued technology education, this logic gives new ways of thinking and acting in technology education, thus opening up possibilities for re-conceptualising past practices and creating future imaginings. I begin by introducing an historical example to give context to the theorisation put forward in the remainder of the chapter.

## Setting the Scene

Back in 1987 Douglas Harper published an anthropological study of everyday technological practices called *Working knowledge. Skill and community in a small shop*. In writing the book, Harper followed the everyday practices of Willie, a mechanical repairman, as he worked on a range of mechanical machines from the town and surrounding farmland where he lived. Willie draws upon his lifetime experience of working with metals and machines, and his intergenerational

[1] In Chap. 2, Keirl introduces the concept of "both-at-once" to conceptualise the Hegelian binary (thesis, antithesis, synthesis), where a binary is considered as "both-at-once" and a "dualism" as an "either-or" phenomenon of two distinct things of mind-matter (noted in Descartes).

blacksmith knowledge (learned as a child from his father) to look after a community's mechanical repair needs. Through photographic and observational journaling, photo elicitation interviews, and ethnographic field notes, a rich description of technological knowledge and skills were documented by Harper. This chapter draws from Harper's study because it represents an historical account of development in technologies in our community in relation to purpose and people. An extract of the dialogue between Willie and Harper follows:

> Willie: "A blacksmith in my father's day, along with shoeing horses, did blacksmith's welds ["You judge that by color. You had to know your temperatures and you had to know your metals to do a blacksmith's weld" (p. 32)]—he built the parts for machinery—well, almost the same principles as I'm doing now, it would be a garage and a blacksmith's shop at the same time...When I'd get home in the evenings from school I'd have to stand on a box to help turn the forge. I was seven years old." (p. 34)

> Harper (1987): "Willie brings his understanding of materials to bear on his work...for knowledge is integral to the method and gives it much of its particular character. Many of the repairs seem unremarkable, such as using a cardboard box to make a gasket or a piece of discarded plastic for a small brace. But the knowledge of materials makes it possible for Willie to use the odds and ends that are in profusion around the shop—to see the value (identified as usefulness) in "junk". Willie's knowledge of the materials helps him understand *why* machines have deteriorated or broken down, and it leads him to see the act of repair as remedying an engineering flaw rather than replacing a part." (p. 47)

Harper's (1987) thesis is that technological knowledge and practices are integrated, systematically understood and contextually bound to purpose and community (see Gumbo, Chap. 4, this volume). Technological activity is therefore closely connected to societal conditions and needs—and these conditions continue to evolve. The historical evolution of technologies can be seen in the way Willie's father's workshop and know-how are transformed over time directly in relation to societal conditions and needs:

> When you bought a car, especially the Model T's, you got a wrench kit with it to adjust your bearings and everything else with. Your bearings were all shimmed. You'd pull a little base pan off the bottom to adjust your bearings. You did most of that yourself. But a lot of people didn't know enough about mechanical work, so they'd take it to the blacksmith's shop. (p. 25)

> This was part of life, Willie says, where "there were not that many things you owned that you didn't make." Even if you'd buy something you couldn't make, preparing it for use often required handwork. (p. 34)

In discussing how Willie both identifies the new knowledge needed in the blacksmith's shop and the growing skills in designing and manufacturing parts, Harper (1987) draws attention to the current *design culture* by capturing this as a 'match and replace' culture, rather than—as observed in Willie's workshop—the systematic and deep *dialectical* knowledge of materials and design that are embedded in purpose and *a valuing of repair*. As noted by Pavlova (2012), the field of technology is now mindful of revisiting these earlier values of repair-maintenance/wise resource use, but in the context of an ecologically aware and concerned citizenship. New thinking, new knowledge, and new technological activity are called for in the context of new societal and global needs for sustainability.

Technology education has an important role to play in re-inventing the present in a changing global context, as it considers the past and the future "both at once" in the context of design and make actions of individuals and collectives in the present.

In Willie's workshop, design and production are also realised "both at once" but for the purpose of maintenance/repair. Here, Willie brings knowledge of the problem and the materials from past activity to both design and produce the needed parts. This process includes "the act of realising thought in object activity, and through activity in the forms of things and events outside consciousness" (Ilyenkov 2003, p. 102). In this dialectical reading of practice and thinking it is possible to see how a futures perspective and the past can emerge within the present moment.

Yet what constitutes the nature of technological knowledge, and what can be defined as the foundational concepts of this field of study, remain contested. That is, the nature of technological knowledge has been—and continues to be—the subject of research attention and debate (e.g., Bjorklund 2008; Dakers 2006; de Vries 2006; Vries and Tamir (1997); Loy and Canning 2012; Stevenson 2008). Williams (2012) states that "Robust debates still exist about the nature of knowledge in technology and the way knowledge empowers technological practice" (p. 170). These debates tend to centre on a dualistic construction of knowledge 'in action' and knowledge 'in the head'. McCormick (2006) suggests that these knowledges have been termed 'procedural knowledge' (know-how-to-do) and 'conceptual knowledge', with the latter referring to relational knowledge. For example, in gearing technology, direction of rotation, change of speed and torque are all interrelated.

McCormick (2006) suggests that depending upon the learning theories drawn upon, different value positions about knowledge will be prioritised:

> Each theory will have a different vision of what it is like to be an expert... Cognitive constructivists are likely to favor a clear set of concepts that are well integrated in the expert's head, and that can be retrieved... Those who support a situated cognition view are likely to talk a different language, as their concern is rather activity, practice and participation and would see expertise in terms of the extent of participation in a community. (pp. 34–35)

The latter sociocultural perspective has caught the attention of many contemporary technology education researchers (e.g., Verillon 2009; Fox-Turnbull, Chap. 6, this volume; Mioduser, Chap. 5, this volume). These new theoretical tools (e.g., Lave and Wenger 1991; Rogoff 1990, 2003) have provided new insights (e.g., Compton et al. 2012; Lemke 2000), which begin to break down the traditional and established practice-theory dichotomy. But what underpins these theories? What are the central concepts that can transform the field of technology education in ways that address the new social conditions and needs of the future? This chapter seeks to draw upon key concepts from *cultural-historical theory* in order to capture and theorise the essence of what I believe constitutes the foundational knowledge of technology education for the future.

While the term *cultural-historical* is used in this chapter, the terms *sociocultural* or *activity theory* could also have been used. I have chosen it because it captures the societal and historical dimensions that are important for understanding past, present and future constructions of the field of technology education, for example, when Willie's story was examined in the context of the current perspective for a more sustainable future. *Cultural-historical* theory is the term used by Russian scholars (past and present) to name the original concepts created by Vygotsky, and developed further by many contemporary scholars (see Stetsenko and Arievitch 2010 for a comprehensive overview).[2] Vygotsky's original theory also inspired the work of Western scholars, such as James Wertsch, Barbara Rogoff, Jean Lave (and many others)—all of whom have influenced the research of technology education. These Western scholars have tended to use the term *sociocultural theory*, a term introduced into English speaking contexts by Jim Wertsch to name the central concepts of Vygotsky. Whilst there are no differences between socio-cultural and cultural-historical when naming the central concepts of Vygotsky, it is how the concepts have evolved in research and been named in subsequent work (e.g., funds of knowledge; communities of practice) that markedly differs.

Stetsenko and Vianna (2009) conceptualise the body of work that has evolved as the *Vygotsky project*. They see "Vygotsky's unique conceptualisation of mind and knowledge as stemming from, participating in, and embodying collaborative social practices, contingent on mediation by cultural tools and, therefore, as encompassing dimensions of knowing and doing in one inseparable blend" (p. 43). Important here is the interrelated system of concepts for realising this holistic worldview for technology education.

---

[2] Within the 'Vygotsky project' has been the work of Leontiev, who further developed the foundational concepts of cultural-historical theory. Leontiev (2005) argued that regardless of the object of the person, the need that arises can be met by collective action, or the application of a tool by others to satisfy a need. At the centre of collective action is the creation of new relationships among people. The central proposition of activity theory is that

> The activity of participants in a joint collective is stimulated by the product of this labor, which initially meets the needs of each of them. . . . Their need is satisfied not by their "intermediate" results, but *by a share in the product of their joint activity* received in every one of them due to the strength of the relationships that emerge in the process of labor that bind them together—that is, social relationships. (pp. 5–6; emphasis added)

Here, motives and a hierarchy of needs become central elements of Activity Theory, where a collective view of practice is foregrounded. Related to Activity Theory is Engestrom's model, which has been influential in technology education research (e.g., Hakkarainen 2009; Leonard and Derry 2011). Cultural-Historical (CH) and Activity Theory (AT) have also been conceptualised under the name cultural-historical activity theory (CHAT). CHAT tends to bring together Vygotsky's system of concepts with Leontiev's conception of *activity* in order to foreground practice more strongly. While there are differences in how each of these theories have evolved, their roots are firmly located in the original theory of L. S. Vygotsky.

What is common to all theories (AT, CHAT, sociocultural, cultural-historical), is the use of dialectical logic for realising "both at once" practice and person in action and thinking. It is a non-reductionist materialist (non-dualistic) conceptualisation of human nature and development that embodies "material social practices of people" (Stetesnko and Arievitch 2010, p. 231). With this reading, the concept of a *Vygotsky project* is useful for thinking about the collection of theories (both original and developed) that underpin thought and practice in technology education.

In working with a holistic view for technology education as introduced through the example of Willie's workshop, this chapter will use the original concepts contained within the collected works of Vygotsky (1930/1997; 1931/1997; 1934/1987), as well as contemporary readings of this literature in relation to empirical studies of technology education where a range of theoretical drivers are used—sociocultural theory, activity theory, and cultural-historical theory. The next section examines the epistemological basis of technology education using cultural-historical concepts in order to lay a new theoretical foundation for technology education.

## The Epistemological Basis of Technology Education

In this resource-deplete field, establishing a set of 'enduring ideas' or concepts for technology education meant that technology teachers had to work from first principles, where a great deal of discussion and negotiation was required for agreement to be reached about what constituted the enduring ideas for technology education. As was also identified by Harper (1987), materials technology was found to be the main topic where students would have "opportunity to see the big picture" and "to articulate its theoretical underpinnings and consequently [work towards the] development of the philosophy that was conducive to a rational epistemology" (Williams 2012, p. 171). The study by Williams highlights the need for an epistemological basis for technology education. Further, it draws attention to what Harper termed the obsolescence design culture, where 'match and replace' rather than a systematic and deep dialectical knowledge of materials, design and embedded purpose—*the project* (see Buntting and Jones, Chap. 10, this volume). In returning to the dialogue between Willie and Harper this idea is taken further: "[To work metal this way] you do one thing at a time, but you're always thinking of the project as a whole and how it's being affected by what you're doing" (p. 34).

*The project* is an important idea for technology education. As can be seen through the work of Harper (1987) and Williams (2012), knowledge is driven by what Verillon (2009) calls "performative action, artefact design and tool use" (p. 181). In other words, the *project* constitutes *knowledge in action* through modelling, building, or testing for a particular outcome. Feedback from testing or modelling validates the design hypothesis, or the diagnosis of a problem. Verillon further argues that in the design process "new knowledge is produced" and this knowledge is always in relation to the "artefact design, fabrication or maintenance" (pp. 180–181) as constituted by *the project*. Thus, a holistic view is realised by technology for both everyday practice—as we see in Willie's mechanical shop—and in education. As Verillon remarks, "What is at stake is not the establishment of 'truth' but the attainment of success and effectiveness: "It works!" is the technician's equivalent of Archimedes' "Eureka!"" (pp. 181–182). But how does knowledge in action speak to a futures orientation for technology education?

De Vries (2006) has written about the nature of technological knowledge, arguing that there are three forms of knowledge: physical, functional, and the relations between these. These are useful for advancing the concept of a futures orientation. The first form of knowledge—physical knowledge—is related to the physical nature of the artefact, that is, being able to know about the artefact, its properties and what it affords. In Willie's workshop this would mean recognising the mechanical device he is going to work on—this constitutes the net knowledge of materials, actions and expected ways of working. This definition can be read more broadly when consideration is given to Harper's (1987) analysis of the physical nature of artefacts in Willie's mechanical shop: "...the *set of tools and materials* is the sum of all previous projects, and that it will be summoned to the task at hand and enlarged once again with the materials left over from the current project" (p. 74, emphasis added).

The second form of knowledge described by de Vries (2006) relates to the functional nature of artefacts, that is, what they are supposed to do—what does a particular form of farm machinery afford for an individual and for a community? The artefact is imbued with historically located and contextually developed needs and purposes. From a cultural-historical perspective, knowledge of the functional use of an artefact is always mediated through others and in relation to its social purpose. Here, a future perspective is embedded in the technology.

According to de Vries (2006) the relations between the physical and the functional nature of artefacts constitute the third form of technological knowledge. He argues that the designer must draw upon a range of disciplines to inform the technological process. In other words, he uses a multidisciplinary and integrated approach for conceptualising the nature of knowledge in design and technology. This definition actively works against a separatist view of knowledge and practice (see also in this volume Stables, Chap. 7; de Vries, Chap. 14) and speaks to a futures orientation. But other concepts are needed to realise this. Unfortunately, researchers in technology education are working in the wake of a dualistic epistemological framing of technology (see Verillon 2009). In contrast, the more recent use of sociocultural, cultural-historical, and activity theory work against rigid 'in the head' and 'in action' constructions of technology education.

In addition, like shadows from the past, the aftermath of research from a cognitive constructivist's perspective has made it difficult for technology education researchers to move the whole field forward. This is because an individualistic perspective of learning and a view of the lone learner make it difficult to see how groups and communities design together and invent artefacts, systems and processes not yet imagined. What is also needed for a futures orientation to technology education, therefore, is a vision of the *technological project as a collective practice*. To achieve this, new concepts are needed for elaborating the foundational basis for future-oriented visions for what constitutes technology education.

## Foundational Cultural-Historical Concepts in Technology Education

While there are many possibilities for what could constitute the foundational knowledge of technology education, this section discusses three interrelated concepts drawn from cultural-historical theory that capture the content of a futures orientation to technology education. They are:

- Tools and signs as cultural practice
- Everyday concepts and technological concepts in technology education
- Imagination and creativity in design and technology education

*Tools and signs* are foundational to technology education because as a dialectic it encompasses the nature and distinctive features of how a futures orientation develops. *Everyday and technological* concepts make visible how the technological project is *both-at-once* conceptual and practical. These concepts allow for a rigorous mechanism to theorise technological activity as bounded by the notion of a *project* where *imagination and creativity* speak directly to how ideas are realised into future artefacts and actions. Together, these concepts provide the content of what can be constituted as a futures perspective of technology education.

### *Tools and Signs as Cultural Practice*

In this section I argue that technology education supports a higher form of psychological functioning and not just a set of practical activities. In this conception, tools and signs constitute a key element of a futures orientation to technology. Technology itself—as materials, practical purpose and design (de Vries 2005)—embodies what it means to be a cultured human being. Human culture as technological activity and thinking does not remain static, but evolves from generation to generation, always as a future enterprise.

It is easier to understand technology as a cultural practice when tool use in non-humans is discussed as a point of contrast: In their longstanding work into the cultural practices of humans, Vygotsky and Luria (1930/1994) argued that what is a central characteristic of humans is not always present in animals' use of tools for future action. They suggest that animals will indeed pick up a stick, fashion it as needed for the present context, and then use it immediately for some purpose directed towards an activity, such as retrieving food. But Vygotsky and Luria (1930/1994) suggest that it is only humans who will fashion a stick for *later use*. Here the tool is directed to some *future activity*. Here the tool is designed for some *future need*. Consequently, it can be argued that the concept of *future* is a distinctly human conceptualisation, and *tools* as a technology aid higher forms of activity and cultural development of societies.

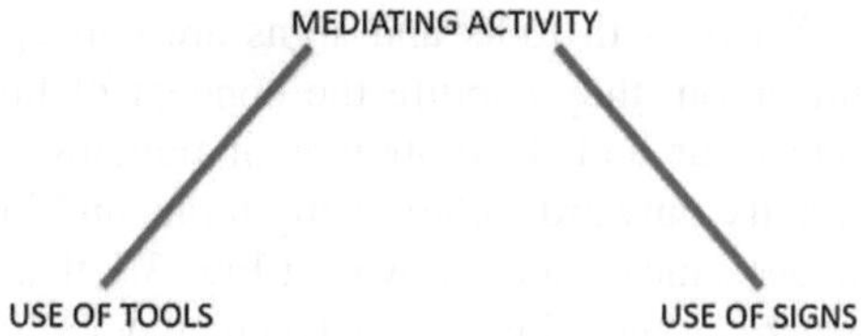


**Fig. 3.1** Vygotsky's conceptualisation of tool, sign and mediation (Adapted from Vygotsky 1931/1997, p. 62)

Vygotsky and Luria (1930/1994) also put forward the claim that humans create signs, such as putting notches in a stick for the aid of memory or for the use of counting in trade. They argued that through the use of signs and tools, human beings enhance their biological capacity. This constitutes an important foundational concept for a futures orientation for technology education. As noted by Verillon (2009), tools and signs "are organizers of activity. They mediate human transformations of the environment" (p. 194), making them a part of cultural rather than natural or biological development of human kind.

Vygotsky's (1931/1997) model of the relationship between tools and signs (see Fig. 3.1) can be used to explain a futures orientation to technology as cultural practice. Here the two types of devices (signs and tools) are shown as "diverging lines of mediating activity" (p. 62). What can be understood from this figure is how tools serve humans towards an activity. The path is outwards, like a spade acting on the environment as a form of "external activity directed toward subjugating nature" (p. 62). The sign, in contrast, directs the action or behaviour of the person (self or another). A form of human-regulation is expressed through the sign. For example, consider the cultural development of the pointing gesture as a sign for infants. Pointing, like the design and use of a spade, is a cultural invention.

Vygotsky (1931/1997) suggested that the initial unsuccessful grasping of an infant directed towards an object (e.g., a spoon) can be conceptualised as an action designed to elicit a behavioural change in a particular social situation. The adult reacts to the outreached arm of the infant as though it were a sign of pointing. It is through another person that the objective of the child is accomplished—and the infant comes to recognise that the action of reaching can in itself change the behaviour of another. Grasping becomes a pointing gesture for direction, and the gesture becomes a sign for directionality. The infant has not only learned a sign (i.e., pointing), but has changed her/his orientation from the object to the person, and through this changed the path of 'child to object' to 'child to adult'.

Consciously knowing and using this sign is representative of a higher mental function, something that is not naturally acquired but rather that is culturally developed through others. Similarly, taking a futures perspective is a learned cultural practice. Verillon (2009) stated that tool-use can be seen as "realising our thoughts in the material world in a way similar to the way words 'realise' our thoughts in speech" (p. 185). Technology acts as mediator and transformer of human culture that is realised collectively for a social purpose. Tools (e.g., language) and signs (e.g., written words) also transform how humans interact in their social and material environments, and when used consciously in technology education, can contribute to new cultural practices and new ways of thinking, such as taking a futures orientation.

When both tools and signs are conceptualised together as part of technology education, they capture the concept of higher mental function or higher cultural behaviour and development of humans. As humans we *both-at-once* direct our activity outwards when using tools, and inwards when using signs. It is this *both-at-once* movement shown in Fig. 3.1 that makes technology education a *cultural practice*. The mind cannot be taken out of the practice equation, and practice cannot be taken out of the mind equation. Fashioning a stick for *future use* is an example of a rudimentary form of a futures orientation to technology education.

As Stetsenko and Vianna (2009) remark, technological knowledge cannot be confined to a separate space of ideas or representations and divorced from the tangible and material "messy" processes of the world of doing things. Doing and thinking cannot be separated into an "inner depiction of outer, mind-independent realities and phenomena that have very little to do with practical (i.e., tangible and material) actions in and on the world" (p. 39). It is through tools used in practice that humans change how they think (e.g., about futures), and it is through thinking about practice that humans change how they do things—otherwise why would a person fashion a stick for later use? Therefore, technology expresses what it means to be human—to engage in cultural practices that do not exist out there in nature to be found or discovered; rather, technologies are human inventions that are passed on from one generation to the next as tools and as signs in activity and thinking.

Technology education systematically supports how humans re-imagine and evolve this cultural knowledge and practice to the next generation, whether it be in classrooms, centres or in the community, as in Willie's workshop. It is the theorisation of this specific form of *pedagogical practice* that society calls technology education. However, technological activity is one of the least theorised human practices. *Theorising technology education as a dialectical relation between tool and sign allows for a more conscious conceptualisation of how a futures orientation develops.*

## *Everyday Concepts and Technological Concepts in Technology Education*

It is argued in this section that theorising a futures perspective of technology education also means taking a collective rather than individual pathway when inventing future-oriented technological concepts. McCormick (2006) suggests that many "tend to see knowledge as an object to be passed around and which will find its way into a learner's head" (p. 31). He argues that this is but one view of the nature of technological knowledge, and a particular view of the mind. He goes on to explain that another view of the nature of technological knowledge is related to the "context in which it has been learned and used ... Individuals who have come to understand ideas in different contexts will have different views of the knowledge they possess" (p. 31). Here, knowledge is not just in the individual mind, but is viewed as being both physically located and socially "intertwined with understanding" (p. 31). This individual and social view of mind sees technological knowledge as situated and mediated through the social prism of human relations rather than as an individual construction (i.e., cognitive constructivism).

In returning to Willie's mechanical repair shop, it is possible to gain insights into how Willie's everyday knowledge is turned into future-oriented technological knowledge for his son Skip as part of a collective cultural enterprise:

> When a car would come in, or a piece of farm machinery—our main work was farm machines when Skip was growing up—he was watching me. I'm hard of hearing so I use a stethoscope on the engines. I'd show him with a stethoscope—how the different sounds—I showed him how to pick out these sounds. I was picking it up with a stethoscope, but he could take the stethoscope off and he could hear the sound easier than I could. (Harper 1987, p. 26)

> I taught him the sounds of different kinds of bad ignition. If a valve is skipping, or burned, it would sound different in the exhaust. Or if you pull the coil wire and turn the engine over you can tell if you've got a burned valve. It'll hiss through the exhaust. If it's burnt real bad you'll get the same thing when it's running at slow speed, on your tractors you can pick up very distinctly because you can idle them right down to about a hundred rpm—you can hear every cylinder just about. If you learn mechanicing on farm equipment the automotive comes 100 times easier. (Harper 1987, p. 27)

Skip was developing valued technological knowledge in the context of both the everyday situation of the engines that had come into the workshop for repair, and through a technological approach where Willie foregrounded *in situ* concepts associated with not only the nature of the different materials (i.e., how metals behave at different temperatures) but also how mechanical structures behave when damaged and how this sounds when parts are moving.

From a cultural-historical perspective, it is possible to conceptualise this as the dialectical relations between everyday concepts and technological concepts. According to Vygotsky (1934/1987), concept formation should be thought about at two levels—the everyday and the abstract (technological).[3] At the everyday level, concepts are learned as a result of interacting directly with the world—developing intuitive understandings of how to do things, such as watching how metals behave when heated. These are important everyday concepts, but the abstract technological concepts underpinning what is being observed may not always be understood.

Vygotsky (1934/1987) argued that these everyday concepts lay the foundations for learning abstract technological concepts. Developing everyday concepts in the context of a person's everyday world is important for living. However, everyday concepts cannot be easily transferred to other contexts. Knowing only about everyday concepts limits a person's thinking to a specific context and reduces opportunities to apply concepts to new situations. As a corollary, Vygotsky (1934/1987) also argued that when someone learns abstract technological concepts

[3] In this chapter, I draw on the concept of everyday and scientific, but rather than stay with the term 'scientific' I substitute the term 'technological' so that the specific abstract concept being considered is located within the field of technology with a futures orientation.

at school away from the context in which they are used (or invented as a form of cultural practice and futures thinking), technological ideas become removed from everyday practice. For instance, learning about the theory of how metals behave when heated without the opportunity to observe this in everyday life makes it difficult for someone to use this knowledge in practice.

In other words, everyday and abstract concept formation are strongly connected to each other. That is, the everyday concepts grounded in the day-to-day life experiences of people create the potential for the development of abstract technological concepts in the context of more formal learning experiences. Similarly, abstract technological concepts prepare the structural formations necessary for the strengthening of everyday concepts in practice (Vygotsky 1934/1987). When someone can both see how metals behave when heated and be given abstract technological knowledge about why different kinds of metals behave as they do when heated for specific contexts and purposes, then learning takes place. Vygotsky (1934/1987) argued that these embedded contexts are important pathways toward abstract technological thought.

These concepts put forward by Vygotsky are linked directly with the situated nature of learning, theorised by Lave and Wenger (1991) as *communities of practice*, by Rogoff (1990) as *apprenticeship in thinking*, and by Moll and Greenberg (1990) as *funds of knowledge*. These latter conceptualisations of pedagogical practices for technology education are useful for highlighting the importance of everyday concepts in practice, particularly where a futures orientation is being conceptualised. But these pedagogical models of situated practice are not new.

Historically-embedded practices for the creation of technological artefacts and processes were established in guilds, where the apprenticeship model flourished. Craft workers became increasingly organised and regulated through guilds, and artefacts became more standardised. This meant that "the old ways with its mélange of hand and theoretical knowledge" (Harper 1987, p. 20) co-existed in guilds and in the community for some time.

Returning to Willie's workshop, it is possible to see through an example this historically developed practical knowledge in action:

> There's what's called "mechanical feel," which is very obvious to those who know what it is, but hard to describe to those who don't; and when you see someone working on a machine who doesn't have it, you tend to suffer with the machine.
>
> The mechanic's feel comes from a deep inner kinesthetic feeling for the elasticity of materials. Some materials, like ceramics, have very little, so that when you thread a porcelain fitting you're very careful not to apply great pressures. Other materials, like steel, have tremendous elasticity, more than rubber, but in a range in which, unless you're working with large mechanical forces, the elasticity isn't apparent. (Harper 1987, p. 118)

*Communities of practice*, *apprenticeship in thinking*, and *funds of knowledge* work with the idea of localised knowledge and a sense of purpose where "the mechanic's feel" for materials constitutes important embedded technological knowledge in practice. However, only when embedded concepts are linked with abstract technological concepts does a futures orientation become possible.

Harper (1987) suggests that the technological activity of Willie lies in contrast to current practices, where there is a "separation from a sense of purpose, separation from self as well as community" and this separation can best be described as "the separation of the worker from the *knowledge* that once guided the work" (p. 21). Cultural-historical theory has sought to re-claim the connectivity, sense of purpose, and concept of *the project* through pedagogical practices in schools that capture "not a laundry list of immutable cultural traits, but rather are historically contingent" (González et al. 2005b, p. 25) knowledge forms that can be consciously considered and re-imagined for the future.

*Fund of knowledge* captures the idea that schools and communities can bring together everyday and abstract technological concepts through the development of "innovations in teaching that draw on the knowledge and skills found in local households" (Moll et al. 2005, p. 71). Examples of funds of knowledge are illustrated in Table 3.1. Moll et al. suggest that "the teacher in these home-based contexts of learning will know the child as a whole person, not merely as a student, taking into account or having knowledge about the multiple spheres of activity

**Table 3.1** A sample of household funds of knowledge

| **Agriculture and mining** | **Material and scientific knowledge** |
|---|---|
| Ranching and farming | Construction |
| Horse riding skills | Carpentry |
| Animal management | Roofing |
| Soil and irrigation systems | Masonry |
| Crop planting | Painting |
| Hunting, tracking, dressing | Design and architecture |
| Mining | Repair |
| Timbering | Airplane |
| Minerals | Automobile |
| Blasting | Tractor |
| Equipment operation and maintenance | House maintenance |
| **Economics** | **Medicine** |
| Business | Contemporary medicines |
| Market values | Drugs |
| Appraising | First aid procedures |
| Renting and selling | Anatomy |
| Loans | Midwifery |
| Labor laws | Folk medicines |
| Building codes | Herbal knowledge |
| Consumer knowledge | Folk cures |
| Accounting | Folk veterinary cures |
| Sales | |
| **Household management** | **Religion** |
| Budgets | Catechism |
| Childcare | Baptism |
| Cooking | Bible studies |
| Appliance repairs | Moral knowledge and ethics |

Adapted from Moll et al. (2005, p. 73)

within which the child is enmeshed" (p. 74). Through knowing the child as a learner of household- or community-based technological knowledge, learning pedagogies become personalised and holistic in relation to the needs of the community or household, as well as in relation to how the learner is connecting with, and understanding holistically, the technological project at hand, that is, gaining a sense of the 'mechanic's feel' for materials.

Rogoff (1990, 2011, 2013) in her early work on *apprenticeship in thinking* and in her more recent work on *intent community participation* makes the point that shared problem solving—with a more active learner in the context of culturally organised activities with more skilled partners—constitutes what is central for learning in an apprenticeship model, where guided participation within routine activities (tacit and explicit communication) "transfers the responsibility for handing skills to novices" (Rogoff 1990, p. 39). Important here is the idea of the value of "including more people than a single expert and single novice; the apprenticeship system often involves a group of novices (peers) who serve as resources for one another in exploring the new domain and aiding and challenging one another" (p. 39).

Hammond (2003), in undertaking a *funds of knowledge* study of a successful technological project—building a traditional house in the school grounds with elders, technology teachers, and the students—noted that "We need to develop educational settings which build on the knowledge which all parties bring to the situation" (p. 39). Here the abstract technological concepts within the school curriculum are understood and enacted through the knowledge of practice that is held within communities and households as valued *funds of knowledge*. González et al. (2005a) argue that pedagogies which actively go beyond the split between school-based knowledge and community-based knowledge position both knowledge construction systems as distributed and not the domain of just an individual.

In drawing upon the concept of everyday and abstract technological concept formation, it becomes possible to make visible how thinking is enacted in practice, and how practice develops the mind. Technological thinking as both-at-once everyday and abstract builds technological concept formation with materials and design, in communities where there is a sense of purpose for the activity. However, in the context of what is unique about the nature of technology education—materials, practical purpose and design—the societal needs that create the development of technologies over time appear to be missing.

In Fig. 3.2, Hedegaard's (2002) conceptualisation of changes in society over time are drawn upon for capturing the central relations between concepts that underpin the nature of technology and design. Figure 3.2 includes the concept of society in a model of technology education where materials, practical purpose, design, and tools and signs are realised together.

In this model the relations between tools and signs, as mediated through person or society as an outward force (tool) or inward movement (sign directing self or others' behaviours), are always centred on purpose and in relation to practice. Specifically, the relations between tools, person and materials (left triangle) act as an outward force. The relations between person, sign and design is an inward-orienting activity. The relations between materials and design (bottom of model) are important because they can be conceptualised in relation to the type of

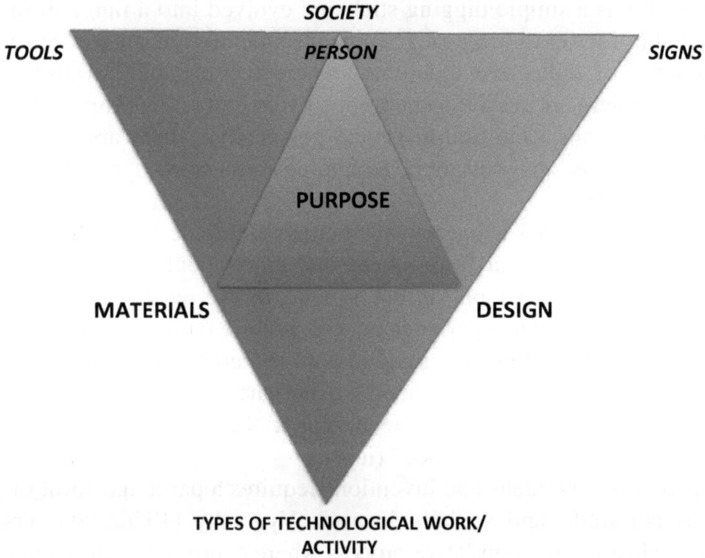


**Fig. 3.2** *Technological unit* as the dynamic relations between materials, practical purpose and design and as changing over time in society

technological work being performed (purpose and division of labour), for example, mechanical engineers or food technologists. Most important, it is the relational dimensions (rather than the nodes) that are important when conceputalising the movement/actions.

As a *technological unit*, Fig. 3.2 seeks to capture the essence of technology education. The model can function as a tool for technology teachers for conceptualising their work, or for analysis of specific pedagogical practices—such as funds of knowledge, apprenticeship in thinking and communities of practice—as pedagogies for technology education. These pedagogical practices explain how historical practices become established in the present community but they do not say very much about inventing what does not yet exist.

Technology is also about imagining and creating what does not yet exist in society—a futures orientation. In the next section, another key characteristic of technology—imagination and creativity—is introduced. An activist stance for technology education is adopted for realising both historical and future conceptualisations as foundational for a cultural-historical theory of technology education.

## *Imagination and Creativity in Design and Technology Education*

Vygotsky (1930/2004), discussing the literature on imagination and creativity, asked how many acts of imagination it took to transform the plough, which commenced life as a simple digging stick and evolved into a range of mechanical devices suitable for small scale (e.g., market gardening) and large scale farming. This is a form of collective creativity, where the original inventors are long forgotten, but where successive generations of inventors continue to improve the original design. From a cultural-historical perspective, there are important links between imagination and design in technology education (see also Spendlove, Chap. 9, this volume).

Vygotsky (1930/2004) suggests that humans imagine what they cannot see, conceptualise what they hear from others, and think about what they have not yet experienced. That is, a person "is not limited to the narrow circle and narrow boundaries of his [sic] own experience but can venture far beyond these boundaries, assimilating, with the help of his imagination, someone else's historical or social experience" (p. 17). The richer a person's experiences, the more possibilities there are for imagining and creating in technology: there is a "mutual dependence between imagination and experience" (ibid).

The creation of new ideas and inventions requires a particular form of imagination that is culturally and socially located. Vygotsky (1930/2004) argued that "Creation is a historical, cumulative process where every succeeding manifestation was determined by the preceding one" (p. 30). Thus, the imagining of an inventor or "even a genius, is also a product of his [sic] time and his environment. His creations arise from needs that were created before him and rest on capacities that also exist outside of him" (p. 30).

A cultural-historical view of design therefore foregrounds imagination as well as creativity: imagination through design is realised in practice as a creation. Put another way, imagination becomes a concrete reality through the making and production process. These imaginings and creations draw upon a rich knowledge of *materials* which constitute the uniqueness of this field, and where cultural-historical theory has shown to be central for the cultural development of humans and for the development of the society in which we live.

The relation between imagination and creativity is dynamic, where technologies not yet invented are imagined and created. Imagination and creativity capture this forward projection. This can be contrasted with some of the other cultural-historical pedagogies that have been suggested to support technology education, such as funds of knowledge, apprenticeship in thinking and communities of practice. However, these pedagogies for technology education on their own give a retrospective account; they do not speak directly to a future orientation. In these pedagogies, it is theorised that novices learn from experts what the experts already know. But what about designing what has not yet been imagined? A futures orientation is possible when imagination and creativity is added to the model shown in Fig. 3.2, as in Fig. 3.3.

Figure 3.3 therefore brings together all the central concepts discussed in this chapter. Materials, design and purpose shaped by societal needs and values are important for fully understanding the unique nature of technology. Tools and signs are mediated through others in the context of particular technological communities, and this creates the conditions necessary for technology to be enacted. Imagination and creativity move design education forward to the future.

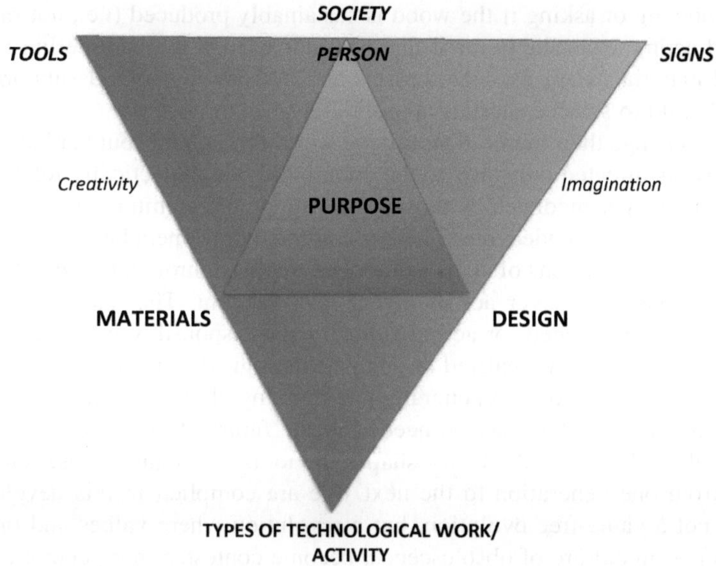


**Fig. 3.3** A futures model of technology and design education

The model shown in Fig. 3.3 contributes to the Vygotsky project for technology education. In this model the central concepts interact together and form the dynamic motion of what constitutes technology education for the future. The model works directly against dualistic views of what has traditionally been discussed as technology education, where each side of the same coin (in the head, in the hand) were theorised separately.

## Conclusion—A futures orientation

A cultural-historical conceptualisation of technology education for future imagining and creating was presented in this chapter. The fundamental concept that was used to discuss development in technology was *tools and signs. Tools* act as an outward movement directed towards changing something in the world (e.g., spade digging), while signs act as an inward movement changing the behaviour of self or others (e.g., pointing). For instance, in technology education students use tools to fashion materials, such as when a preschool child hammers a nail through a bottle top into a piece of off-cut pine to represent a propeller on a plane, or a secondary school student uses a lathe to turn a piece of soft timber into a small bowl. Both child and student direct their efforts towards imagining and creating something with the aid of a tool—a hammer (plane) or lathe (bowl). Both are changing something that exists in the world—a piece of timber. In this example, materials can also act as a sign through being given particular meaning in society, such as a secondary student knowing or asking if the wood is sustainably produced (i.e., not rainforest trees) and acting accordingly, or if the materials used at the hammering table are re-cycled materials (i.e., used bottle tops and not new ones) and supporting the preschool child to select materials appropriately.

Humans change their material and social world through this outward and inward activity. Tools directed outwards to the community are dialectically related to the inward path of signs mediated by those from within our community who give value and knowledge to this endeavour. There is continual movement between the inward and the outward directions of signs and tools as realised through the social prism of human relations. We never act or invent in a vacuum. The dialectical relation creates the tension needed for acting ethically and responsibly in relation to what is imagined, designed and realised through technological activity. This tension also creates the discord needed to generate new thinking about what has not yet been invented, and about what may be needed in the future. The tension—as realised through values, beliefs and ethics—shapes the tools and signs we use, create and pass on from one generation to the next. We are complicit in this development, which is not a value-free evolution, but a revolution where values and practices, such as a design culture of obsolescence, become contested intergenerationally as new circumstances, needs and demands present themselves. Technology education has a major role to play in making visible how tools and signs as drivers of technological development are always realised through the social prism of human relations.

The dialectical concept of *everyday concepts and abstract technological concepts* was introduced to show the 'both-at-once' relations between 'in the hand' and 'in the head' and to push against the assumed dualism that has pursued technology into the education systems found in many countries around the world. Through the lens of cultural-historical theory, it was shown that technology encompasses a unique system of concepts that cannot be separated into a dualism of 'the hand' and 'the mind'.

In addition, cultural-historical theory shows technology to be a valued form of cultural knowledge and practice (Vygotsky 1930/1994) that has mediated individuals and collectives and transformed societies over time. Being blind to this knowledge base is like ignoring species diversity within an ecological system. The cultural and historical ecology of societies is realised through the knowledge base and values-oriented activity of technology. Technological learning as theorised through the dialectical relations between everyday and abstract technological concept formation has a central role to play in developing innovative pedagogies where practice and concepts are 'both-at-once' supporting the new thinking needed as communities grapple with issues of sustainability for the survival of the planet, and matters of cultural validity in knowledge generation for global citizenship.

The third dialectical concept introduced in this chapter was *imagination and creativity*. This concept was put forward in order to understand design in technology education as being futures oriented—thinking and creating into the future. Design in technology education has a significant contribution to make in developing the imaginative and creative thinkers needed for realising future possibilities for societies who now trade in ideas and innovations, while at the same time taking cognisance of the central values of ethics and sustainability enacted in Willie's small mechanical workshop where there was no separation from a sense of purpose, or a separation from self or community. Instead, technological activity was embedded in a *deep knowledge* of purpose and community where the actions of individuals were immediately felt (see also Gumbo, Chap. 4, this volume). Together, these constituted the idea of *the project.*

Technology education develops the individual to imagine and create things not yet designed and produced and, through this, technology education contributes to the technological development of society. As such, technology education has both an ethical and moral responsibility to support imaginings that sustain people and communities in harmony and for the well being of the broader ecological and social environment. Together, the cultural-historical concepts introduced in this chapter help to theorise technology education as a higher form of psychological development of the student—a student who can contribute to the continued evolution and revolution of society.

In projecting to the future, technology education must be conceptualised both pedagogically and morally as a social and ethical practice. When technology education is theorised as a mind-body split, technology becomes objectified and sits outside of human practice. However, when technology is theorised using cultural-historical concepts, it becomes subjectified, meaning that technology sits within the context of cultural practice. The three interrelated concepts of sign-tool, everyday-technological, and imagination-creativity constitute the foundations of a cultural-historical theory of technology education. Such a reimagining of technology education as a cultural practice brings to the fore a futures orientation to thinking and practice.

## References


Bjorklund, L. (2008). The repertory grid technique. Making tacit knowledge explicit: Assessing creative work and problems solving skills. In H. Middleton (Ed.), *Researching technology education. Methods and techniques* (pp. 46–69). Dordrecht: Sense.

Compton, V., Compton, A., & Patterson, M. (2012). Student understanding of the relationship between fit for purpose and good design: Does it matter for technological literacy? In H. Middleton (Ed.), *Explorations of best practice in technology, design and engineering education* (Vol. 1). Proceedings of the 7th Biennial International Technology Education Research Conference (pp. 68–78). Brisbane: Griffith Institute for Educational Research, Griffith University.

Dakers, J. R. (Ed.). (2006). *Defining technological literacy. Towards an epistemological framework*. New York: Palgrave MacMillan.

De Vries, M. J. (2005). *Teaching about technology. An introduction to the philosophy of technology for non-philosophers*. Dordrecht: Springer.

De Vries, M. J. (2006). Technological knowledge and artifacts: An analytical view. In J. R. Dakers (Ed.), *Defining technological literacy. Towards an epistemological framework* (pp. 17–30). New York: Palgrave MacMillan.

De Vries, M. J., & Tamir, A. (Eds.). (1997). *Shaping concepts of technology: From philosophical perspectives to mental images*. Dordrecht: Kluwer.

González, N., Andrade, R., Civil, M., & Moll, L. (2005a). Funds of distributed knowledge. In N. González, L. Moll, & C. Amanti (Eds.), *Funds of knowledge. Theorizing practices in households, communities, and classrooms* (pp. 257–271). Mahwah: Lawrence Erlbaum Associates.

González, N., Moll, L., & Amanti, C. (Eds.). (2005b). *Funds of knowledge. Theorizing practices in households, communities, and classrooms*. Mahwah: Lawrence Erlbaum Associates.

Hakkarainen, K. (2009). A knowledge-practice perspective on technology-mediated learning. *Computer Supported Collaborative Learning, 4*, 213–231.

Hammond, L. A. (2003). Building houses, building lives. *Mind, Culture, and Activity, 10*(1), 26–41.

Harper, D. (1987). *Working knowledge. Skill and community in a small shop*. Chicago: The University of Chicago.

Hedegaard, M. (2002). *Learning and child development: A cultural–historical study*. Aarhus: Aarhus University.

Ilyenkov, E. V. (2003). *The ideal in human activity. A selection of essays by Evald Vasilyevich Ilyenkov*. Pacifica: Marxist Internet Archives.

Lave, J., & Wenger, E. (1991). *Situated learning: Legitimate peripheral participation*. Cambridge: Cambridge University.

Lemke, J. L. (2000). Across the scales of time: Artifacts, activities, and meanings in ecosocial systems. *Mind, Culture, and Activity, 7*(4), 273–290.

Leonard, M. J., & Derry, S. J. (2011). *What's the science behind it? The interaction of engineering and science goals, knowledge, and practice in a design-based science activity* (Working paper No. 2011–5). Retrieved from University of Wisconsin–Madison, Wisconsin Center for Education Research website: http://www.wcer.wisc.edu/publications/workingPapers/papers.php

Leontiev, A. N. (2005). Lecture 13. Language and consciousness. *Journal of Russian and East European Psychology, 43*(5), 5–13. (Lecture was given on 20th December 1973, Moscow University)

Loy, J., & Canning, S. (2012). Changing the emphasis of the learning through making in technology education. In H. Middleton (Ed.), *Explorations of best practice in technology, design and engineering education* (Vol. 2). Proceedings of the 7th Biennial International Technology Education Research Conference (pp. 19–24). Brisbane: Griffith Institute for Educational Research, Griffith University.

McCormick, R. (2006). Technology and knowledge: Contributions from learning theories. In J. R. Dakers (Ed.), *Defining technological literacy. Towards an epistemological framework* (pp. 31–47). New York: Palgrave MacMillan.

Moll, L. C., & Greenberg, J. B. (1990). Creating zones of possibilities: Combining social contexts for instruction. In L. C. Moll (Ed.), *Vygotsky and education: Instructional implications and applications of sociohistorical psychology* (pp. 319–348). Cambridge: Cambridge University.

Moll, L., Amanti, C., Neff, C., & González, N. (2005). Funds of knowledge for teaching: Using a qualitative approach to connect homes and classrooms. In N. González, L. Moll, & C. Amanti (Eds.), *Funds of knowledge. Theorizing practices in households, communities, and classrooms* (pp. 71–87). Mahwah: Lawrence Erlbaum Associates.

Pavlova, M. (2012). Generic green skills: Can they be addressed through technology? In H. Middleton (Ed.), *Explorations of best practice in technology, design and engineering education* (Vol. 2). Proceedings of the 7th Biennial international technology education research conference (pp. 49–57). Brisbane: Griffith Institute for Educational Research, Griffith University.

Rogoff, B. (1990). *Apprenticeship in thinking: Cognitive development in social context*. New York: Oxford University.

Rogoff, B. (2003). *The cultural nature of human development*. New York: Oxford University.

Rogoff, B. (2011). *Developing destinies. A Mayan midwife and town*. Oxford: Oxford University.

Rogoff, B. (2013). *Intent community participation*. Retrieved from www.intentcommunityparticipation.net

Stetsenko, A., & Arievitch, I. M. (2010). Cultural-historical activity theory. Foundational world-view, major principles, and the relevance of sociocultural context. In S. R. Kirschner & J. Martin (Eds.), *The sociocultural turning psychology. The contextual emergence of mind and self* (pp. 231–252). New York: Cambridge University.

Stetsenko, A., & Vianna, E. (2009). Bridging developmental theory and educational practice. Lessons from the Vygotskian project. In O. A. Barbarin & B. H. Wasik (Eds.), *Handbook of child development and early education. Research to practice* (pp. 38–54). New York: Guilford.

Stevenson, J. (2008). Capturing knowledge and activity. In H. Middleton (Ed.), *Researching technology education. Methods and techniques* (pp. 155–171). Dordrecht: Sense.

Verillon, P. (2009). Tools and concepts in technological development. In A. Jones & M. de Vries (Eds.), *International handbook of research and development in technology education* (pp. 175–197). Dordrecht: Sense.

Vygotsky, L. S. (1930/1997). The problem of the cultural development of the child. In R. Van Der Veer & J. Valsiner (Eds.), *The Vygotsky reader* (pp. 57–72). Oxford: Blackwell.

Vygotsky, L. S. (1930/2004). Imagination and creativity in childhood. *Journal of Russian and East European Psychology*, *42*(1), 7–97.

Vygotsky, L. S. (1931/1997). The history of the development of higher mental functions. In L. S. Vygotsky & R. W. Rieber (Eds.), *The collected works of L.S. Vygotsky, Vol. 4* (M. H Hall, Trans.). New York: Plenum.

Vygotsky, L. S. (1934/1987). Problems of general psychology. In R. W. Rieber & A. S. Carton (Eds.), *The collected work of L.S. Vygotsky, Vol. 1* (J. E. Knox & C. B. Stevens, Trans.). New York: Plenum.

Vygotsky, L. S. & Luria, A. (1930/1994). Tool and symbol in child development. In R. Van Der Veer & J. Valsiner (Eds.), *The Vygotsky reader* (pp. 99–174). Oxford: Blackwell.

Williams, P. J. (2012). Technology teachers PCK: The need for a conceptual revision. In H. Middleton (Ed.), *Explorations of best practice in technology, design and engineering education* (Vol. 2). Proceedings of the 7th Biennial International Technology Education Research Conference (pp. 165–179). Griffith Institute for Educational Research, Griffith University.

# Chapter 4
# Indigenous Technology in Technology Education Curricula and Teaching

**Mishack T. Gumbo**

The premise of this chapter is that indigenous technologies have a place in Technology Education, and a case is made for the integration of indigenous technology into Technology Education curricula. The potential outcomes are profound—students from both Western and indigenous cultures who are empowered to participate in the development and critique of technologies from multiple perspectives, widened scope for community participation in teaching and learning, and enhanced collective participation of the custodians of indigenous and Western knowledge systems. The implications of such an approach encompass content, materials and equipment, pedagogies and assessment. First, curriculum developers and teachers need to understand and commit to the value of an integrated approach.

## Introduction

> ....I was travelling in Africa and—whilst reading a local newspaper in a Zambian airport lounge—came across an intriguing advert for a university research post in 'indigenous knowledge'. The more I read about it, the more intriguing it became and the more questions it raised for me, the most central being what is 'indigenous knowledge'? (Kimbell 2008, p. 8)

During my scholarship review for this chapter my eyes caught the short article entitled *Indigenous knowledge, know-how, and Design & Technology* from which I quoted the above excerpt. Knowing Richard Kimbell as an expert in Design and Technology, I was very keen to read about his views on indigenous knowledge from a Design & Technology perspective. My reading of this article and other literature

M.T. Gumbo (✉)
Department of Science and Technology Education, University of South Africa, Pretoria, South Africa
e-mail: gumbomt@unisa.ac.za



PJ. Williams et al. (eds.), *The Future of Technology Education*, Contemporary Issues in Technology Education, DOI 10.1007/978-981-287-170-1_4

identified a gap. The gap is that, despite extensive scholarship in indigenous knowledge systems (IKS) and indigenous technology in particular (e.g., Emeagwali 2003; Green 2008; Msila 2009; Nakpodia 2010; Odora Hoppers 2002; Vandeleur 2010; Zulu 2006), there has been little discussion on the integration of indigenous technology in the Technology Education curriculum. In addition, there are numerous reports that "Traditional approaches to learning . . . have not mobilized indigenous knowledge and expertise among many people" (Carvallo 2000, p. 769). Instead, institutions of learning have largely proven resistant to change, not accommodating indigenous technological forms and contributions. For instance, the findings of a comparative study between South Africa and New Zealand—two countries with significant indigenous populations—revealed technology lessons that are devoid of indigenous technology issues. Further, technology teachers seemed not to understand the concept (Gumbo and Williams 2012; Williams and Gumbo 2011). As a consequence, there seems to be too much emphasis on teaching western technological knowledge instead of balancing it with indigenous technological knowledge.

As a Technology Education specialist in South Africa, where the African population is nearly 80 % and the white population less than 10 %, I find this state of affairs concerning—particularly given the collaborative and cooperative pedagogical approaches, and diverse forms of technology that Technology Education presents; as well as the socio-cultural and multicultural realities that learners represent in Technology Education classes. At stake in this chapter is the need to integrate indigenous technology in the Technology Education curriculum, taking into account the ideas of Marilyn Fleer (Chap. 3), David Mioduser (Chap. 5) and Wendy Fox-Turnbull (Chap. 6), who all contribute to this book from socio-cultural, historical-cultural and multicultural perspectives.

The contention that IKS have been deliberately marginalised puts the blame squarely on the west's colonial practices, pushing the poles of IKS and western knowledge systems (WKS) further apart rather than bringing them closer. What complicates this matter even further is the polarisation of the western world (predominantly characterised by 'modern' technology) and southern world (predominantly characterised by indigenous technology). This polarisation has even translated into regional differences expressed through rural and urban contexts—the prevalence of indigenous knowledge diminishes into urban contexts. Hence, learners, who are much implicated in the discussions in this chapter, are now divided in two worlds. This challenges attempts to embed indigenous perspectives in the curriculum. However, the so-called modernised learner still shares ties with his or her indigenous milieu through occasional interaction with it. It is from this premise that I call for a paradigm shift in an effort to integrate (Marinova et al. 2010) indigenous technology in the Technology Education curriculum to enhance the collective and progressive participation of the custodians of both WKS *and* IKS. Further, I argue that both indigenous technology and western technology have potential to find a common course into the future (Omolewa 2007; Seemann 2000). Thus, my task in this chapter is to map out the educational parameters that characterise indigenous knowledge/technology in relation to the ideological orientation of western education. This enables me to argue for the integration of indigenous technology in the Technology Education curriculum by offering reasons for its inclusion. Finally, my route brings me to a point of suggesting a Technology Education curriculum that integrates indigenous technology, and reflecting on the challenges that this poses to teachers.

## A Case for Integrating Indigenous Technology into the Technology Education Curriculum

### *Defining Indigeneity, Indigenous Knowledge, Culture, and Indigenous Technology*

I begin this section by briefly defining key terms used in this chapter, including indigeneity, indigenous knowledge, culture, and indigenous technology. Indigeneity (being indigenous) means the root of things or something that is natural/inborn to a specific context or culture (Odora Hoppers 2002; Van Wyk 2002). This definition does not come as a surprise when considering my South African context. I want to use a scenario of animals and plants, and languages, to illustrate the meaning of indigeneity: When talking about indigenous fauna or flora, it is expected that reference is made to animal or plant species that originate in South Africa compared to exogenous species. In institutions of learning and government institutions, a distinction is drawn between 'western' and 'indigenous' through indigenous languages. In South Africa, these refer to black cultures' languages, such as Tswana, Zulu and Shangaan, compared to the predominant languages—English and Afrikaans—which have European roots. In certain academic institutions in South Africa there is even a Department of African (Indigenous) Languages. For purposes of this chapter I use the term indigeneity to refer to the people indigenous to a specific context or region, and their knowledge systems, which include technology.

Let me come back to Richard Kimbell (2008) being intrigued by the concept of indigenous knowledge. In his quest to find out more about it, he scratched for the answer. He states that:

> ... on the surface the question is easy to answer through examples ... The bushmen of the Kalahari know how to find water in their parched landscape by reading the signs that they see in the environment but that others do not observe. This knowledge is central to their survival, and is passed down from generation to generation through an oral and experiential tradition. (p. 8)

But, perhaps providing one example only may not satisfy the broad meaning of the term. Thus, indigenous knowledge is that knowledge that is held and used by people who identify themselves as indigenous to a place based on a combination of cultural distinctiveness and prior territorial occupancy relative to a more recently-arrived population with its own distinct and subsequently dominant culture (Mugabe n.d.). Indigenous knowledge therefore has to do with the complex set of activities, values, beliefs and practices, has evolved cumulatively over time, and is active among communities and groups who are its practitioners (Owuor 2007).

Owour (2007), in a study about indigenous Kenyans, observes that where formal education has had insignificant impact, oral art remains the most important means of transmitting knowledge and skills as a way of maintaining societal continuity from one generation to the next. Owour cites an example in this regard, stating that during initiation into adulthood among the Kikuyu, Maasai, Luhya, and Kalenjin communities, the elders prepare youths for their transitional roles and responsibilities in adulthood. This has implications for indigenous education in the Kenyan context. For example, Owour writes that the methods used in indigenous education are aimed at integrating character building, intellectual training, manual activities, and physical education. Specific trade skills are learnt through apprenticeship and youths' observations of the practices modelled by adults or trainers. There are therefore two kinds of knowledge, each learnt in particular ways. One is a specialised knowledge, such as indigenous medicine and spirituality. Specific members of the family are identified as custodians of the knowledge and mentored through exposure to the practice by those who are specialised in the field from the family or clan. The other is experiential knowledge, which is always acquired through personal exploration and practicality based on everyday lived experiences. Case-based research by Marchand (2008) draws from different contexts—minaret builders in Yemen, mud masons in Mali and fine-woodworkers in London. These cases demonstrate the role played by elders from indigenous craft and apprentice perspectives in knowledge and skills transfer to the young (inexperienced) ones on-site. Education and training happens through imparting innate knowledge, demonstrations and observations, while ensuring social, religious, and so forth cohesion.

These kinds of knowledge raise issues of epistemology—what and whose knowledge gets legitimised in the school curriculum? Since knowledge is power-, politics- and culture-bound (Apple 2000; Gegeo and Watson-Gegeo 2002), it is not value-free. It is contested terrain, in the context of this chapter between west and indigenous. The western approach to science and technology marginalises indigenous forms of science and technology (Emeagwali 2003), accusing them of being oral and devoid of proof (Castiano 2011). Such 'alternative forms' of knowledge and knowing have been unnecessarily restricted in the knowledge and curriculum domains. As a result, "There is a need to develop ..... the epistemological basis to technology studies in schooling. Without this depth of understanding the field of technology education has little hope of meeting its potential ...." (Seemann 2000, p. 1).

From the types of knowledge referred to above, Owour (2007) deduces that indigenous education involves the expertise of multiple teachers, given the multiple natures of roles and responsibilities in life through which young people are mentored and guided. This is the type of approach to education that I also grew up knowing, and to this day it is well known among my community members that "Ngwana o godisiwa le go rutiwa ke setshaba" (Tswana for: "It takes a community to raise and educate a child").

Culture is the way of life of a social group that includes actions, values and beliefs that can be communicated, with necessary modifications, from one generation to the next. Nakpodia (2010) writes that culture is learnt; dynamic because it varies from one society to another; and a complex whole which includes knowledge, beliefs, art, morals, law and customs. Thus, one of the major functions of education is to transmit culture—understandings of technology, history, literature, philosophy, science, etc.—to the young (Lawton 1982). In line with this view, there are three important implications for education: culture is educationally transmitted, learnt and shared. According to Seemann (2000), cognitive activity is inseparable from its cultural milieu, and every society educates the younger generation as a means of passing down its socio-cultural attributes that guide what a child learns and becomes. This means that Science and Technology Education, too, are human enterprises that involve the transmission of cultural heritage (see also Fleer, Chap. 3) and should take into account IKS. Such an approach would align well with Technology Education—Seemann (2000) argues that indigenous communities' approaches to problem solving are holistic, characterised by the intertwining of culture, technology and environment.

Indigenous technology is a body of knowledge, developed by a culture, that provides methods or means to control the environment, extract resources, produce goods and services, and improve the quality of life (Cheek 1992). Indigenous technology includes technologies such as looms; textile, jewellery and brass-work manufacture; and technological knowledge in agriculture, fishing, forestry, resource exploitation, atmospheric management techniques, knowledge transmission systems, architecture, medicine and pharmacy (Odora Hoppers 1998). Obikeze (2011) describes indigenous technology in terms of tangible devices (knives, fishing nets, machines, bombs, electronic devices, and so on) and intangible devices (songs, jokes, ideas, skills, methodologies, organisations, and so on).

Bearing in mind the above definitions, I claim that technology is inherently cultural (Custer 1995; McCade and Weymer 1996; Potgieter 1998). This is consistent with Custer (1995), who argues that the notion of technology as artefacts extends beyond physical objects: "artefacts are seen as wonderful and diverse cultural expressions" (p. 223). When learners engage in a design and make task in the learning of Technology, reference is made to an artefact as a product that they should ultimately make. I propose that in this process, learners should be exposed not only to the western notions of design and making, but to indigenous forms as well. I critically reflect on the curriculum that I received in my formal education, which did not include indigenous knowledge forms. Even when my specialisation in Technology Education started, when I was studying for my Master's degree in Education, I was confronted by readings on western perspectives by western writers and technologists only. There were no attempts by the designers of the course to integrate indigenous technological perspectives, yet these are manifold.

- Food technology examples include dehydrated granular food products, which involve fermentation, frying and dejuicing; or products such as sorghum, maize, or other cereal fermented and made into alcoholic beverages; various types of cereal-based flour, pulverised tubers of various kinds and a wide variety of vegetable-based soups (Okagbue in Emeagwali 2003).

- Examples of metallurgical technology include carbon steel production 1,500–2,000 years ago on the western shores of Lake Ukerewe in Tanzania; copper smelting developed independently in West Africa around 900 AD (ATPS 2010; Emeagwali 2003; Sertima 1983).
- Then there is astronomical technology, like a stone astronomical observatory created in Kenya on the edge of the Lake of Turkana (Adams 1983; Tedla 1995). ATPS (2010) reveals tools technology of bone tools and blades in Southern and Eastern Africa 90,000–60,000 BC; use of iron smelting and forging for tools which appeared in Africa around 1200 BC.
- Construction of Great Zimbabwe more than 800 years ago and discoveries of ancient mines; ivory, gold, sacred birds of soapstone, divination bowls and dishes; sophisticated guns made from iron; currencies consisting of gold and brass made from metal coinage on East African and western coasts; discovery of Khoi ceramic pottery in Mpumalanga and of iron production in Cameroon; discovery of golden artefacts like rhino and bracelets at Mapungubwe can be categorised as architecture and engineering technology (Hall 1996; Emeagwali 2003; Asante and Asante 1983; Maluleka et al. 2006; Orevbu 1997; Tedla 1995).
- Transport technology includes construction of watercrafts for jungle canoes and dugouts from reed and wood, with cooking facilities (Sertima 1983).
- Agriculture technology includes cultivation and harvesting of barley, cowpea, millet, sorghum, yam, coffee and cocoa; the use of different cropping systems; and domestication of cattle (Atte 1992; Rowlands and Warnier 1996; Sertima 1983).
- Medical technology examples include aspirin; use of bark of salix capensis to treat musculoskeletal; rootbark annona senegalensis to treat cancer; herbs to treat retarded labour, malaria fever, rheumatism, snakebite, etc. (Emeagwali 2003; Jonathan 1996; Sertima 1983).
- Lastly, examples of communications technology include drumming scripts used to relay news over great distances and for celebratory music and dance (Tedla 1995).

Following the definition of indigenous knowledge above, I hold a view that technology resides in a knowledge domain, which is why I maintain that indigenous technology is part of IKS. According to Battiste (2002), indigenous knowledge comprises the complex set of technologies developed and sustained by indigenous civilisations. In Robyn's (2002) study on Native American Indians, Grenier finds a reason to see a synergy between indigenous knowledge and indigenous technology. Grenier claims in this regard:

> Since the very survival of Native peoples depended on their being able to utilize knowledge in balance with the natural environment, one could make the argument that indigenous knowledge is technology. (in Robyn 2002, p. 199)

Kimbell (2008) relates this synergy and sees it as befitting Design & Technology:

> Most of what might be termed the 'indigenous' knowledge that I came across in Africa—as well as most of the references to it that I have subsequently read—relate to practical knowledge; the kinds of know-how that make life live-able in the local situation. It's about growing or hunting for food, building shelters, or transport systems, developing tools and apparatus and systems. In short, indigenous knowledge is typically design and technology knowledge, which is 'know-how' rather than 'know-that'. (p. 9)

Kimbell's (2008) view was informed by his observation of an apt example of this know-how while in Zambia, at a beach construction site for dhows, the traditional Red Sea/Indian Ocean sailing craft with its characteristic triangular (lateen) sail:

> I watched as a big-ish 25 ft dhow was being constructed. Raw materials (typically branches/trunks of teak) were being selected, shaped and fixed, all by hand and without a single drawing. The builders 'knew' about the strength of the timber and how to shape and fix it, and they looked for particular pieces to do special jobs within the construction. (p. 9)

In this observation, Kimbell reflected on tacit knowledge. He realised that new members of the building group were being progressively inducted through participation in the 'mysteries' of the trade of building sailing craft. He concluded that it is not so much personal knowledge but participatory knowledge being demonstrated in this enterprise.

With the realities of indigenous technology and the possible benefits that it can bring to the entire humanity, I suggest that ignoring the importance of indigenous technology in Technology Education curriculum should be tolerated no longer. I thus concur with Rains (1999), who argues:

> When we fail to include sophisticated understandings of indigenous knowledge in the curriculum, when we fail to teach well, when we fall prey to historical amnesia, when we buy into the contemporary intellectual authority, we are granting jurisdiction over complacency within the status quo. (p. 328)

According to Custer (1995), a more balanced technological perspective is to begin with imagination and culture, and then consider and appreciate the wonderful diversity that has been created. Instead of form (machine, tool, artwork, score of music, etc.) being a distinguishing criterion, the emphasis should be on the ways in which the values, priorities and needs of various cultures take form through the creative energy of their people. It follows, then, that extending the learning of technology to include indigenous forms will enrich and expand the scope of learners' concepts of technology. Further, the fact that physical or technological artefacts cannot be separated from culture suggests that if the aim of the curriculum is to be culturally sensitive, then it must integrate indigenous technology. This leads me to explore the differences between IKS and WKS so that later on I will be able to argue for the inclusion of indigenous technology in the Technology Education curriculum.

## Differences Between IKS and WKS

Recognising the differences between IKS and WKS supports efforts to bring the two together. Hence, my aim with this section is not to promote a dichotomous stance, but to acknowledge the very real differences so that we can begin to re-orientate our efforts to explore lines of convergence between them and facilitate integration.

## *IKS*

Tedla (1995) explains the philosophy informing IKS-based education and characterises it in terms of academic excellence, spiritual development, community development and physical fitness and health. Regarding academic excellence, education content reflects the reality and needs of indigenous people; combines abstract learning with practical learning, and book learning with experiential learning; involves the entire community in the educational process, with local communities actively participating in shaping their educational destiny; produces communities of scholars and learners who are indigenous-oriented; forms study groups that focus on expanding knowledge about indigenous people and creating new ways of solving problems; and avoids purely individualistic and competitive approaches to education, for education is communal by its nature in order to enable one to live in harmony as a contributing member of the local and world communities. From an African community perspective, such education is informed by principles of ubuntu. Ubuntu is better understood through the values that define a communal society—group solidarity, conformity, compassion, respect, human dignity, a humanistic orientation and collective unity (Mokgoro 1997). The community defines the person as a person. In this sense, a definition of a person is expressed in terms of being a biological relative of a broad family. This explains the extended family ties of indigenous communities, which resonate to some extent with the concept of community of practice. The interdependency of such communities is described by Barab et al. (2002) as a group of individuals who are socially interdependent and who share mutually defined practices, beliefs, and understandings over an extended time frame in the pursuit of a shared enterprise.

Spiritual development is based on the notion that spirituality permeates all aspects of life of indigenous peoples. The held view is that this is a sacred world, and reverence for life dictates that everyone acts 'right' by each other. This means that spiritual education should include the development of a character that respects life, that is, that preserves, nurtures and affirms life; takes care of elders, orphans and the weak; learns from the wisdom of elders; is generous, honest, just and diligent; strives for excellence in everything; fights oppression with a clear heart and strong spirit; and believes in the community, in self, in life.

Community development suggests education that recognises the inseparability between and complementarity of the individual and community, inculcates respect for elders, teaches building and maintaining strong family ties, produces individuals that participate in the political, economic and educational life of traditional communities, and takes full control of the education of indigenous children and ensures their mastery of many practical skills—indigenous crafts, technologies and medicine. Such education changes the content that is taught to indigenous children so that it reflects the values and needs of their community, teaches all that is positive in traditional leadership and governance, strives to minimise individualism and competitiveness, produces people who participate voluntary community service on an on-going basis, and recognises that adults and elders have a duty to mentor the young.

Regarding physical fitness and health, education teaches preventative health measures, re-orientates towards food that is currently proven to promote good health, encourages learning from indigenous women in agriculture about naturally cooked food as a community involvement, and incorporates learning activities which include indigenous methods of teaching and games.

From these characteristics of IKS, it can be seen that an education relevant to indigenous learners should be holistic and informed by elements that emanate from the indigenous philosophical stance relating to academic excellence, spiritual development, community development and physical fitness and health. Most importantly, such education should recognise the role that indigenous elders can play in teaching the young ones.

## *WKS*

The ideological stance of WKS-based education is explained by Coelho (1998) as a context where students bring stereotypical attitudes to the school formed by forces outside the school, like the family, media and interactions in the community. The school may unintentionally reinforce these attitudes through the curriculum it offers to students. Students of the dominant culture are harmed by a curriculum that represents, affirms, and celebrates only their cultural background and experience. Students whose cultural backgrounds are not validated by the curriculum receive the implicit message that their cultures are not worthy of study, and that people of these cultures have achieved little and contributed nothing to human history. Resultant impacts on such students include poverty (when they cannot find jobs because they are trained in redundant fields in the job market); unequal distribution of educational resources (e.g., their education receives far less funding compared to their western counterparts); lack of opportunities to learn (due to their marginalisation, some cannot afford education due to their poor backgrounds); difficulty engaging with the language of instruction (they predominantly receive their education through the medium of English); low teacher expectations (teachers tend to engage them minimally in learning activities as they view them as teachers think they cannot contribute much); and a mismatch between teaching and learning styles (teachers' pedagogical strategies do not accommodate these students' learning styles).

Western curricula tend to focus on the perspectives, experiences, achievements, contributions, inventions, discoveries, creations, beliefs and daily life activities of people of European ancestry, and may even distort or omit those of other groups. Such curricula are limited in their selection of knowledge that students should learn, providing them with a biased view of the world because they, at worst:

- study literature by Europeans but read little or no literature from other cultures;
- learn and know a lot about the arts of Europe but remain ignorant of the artistic forms and creations of the rest of the world;

- learn in social studies that the family usually consists of two parents and two children, and that people in the 'Third World' are poor because they have too many children;
- in the history class, they learn about the arrival of the European explorers and colonists from a European perspective;
- learn in mathematics, science and technology little of the involvement of other cultures in those fields;
- graduate with limited proficiency in any other language; and
- are taught mostly by teachers who are members of the dominant culture.

More traditional home-based education seems to have worked better when the provision of education was the responsibility of parents and families and later, for some, the church (Seemann 2000). This is important for Technology Education since it interfaces with other subject areas like Mathematics, Science and Arts—something that speaks well to the holistic philosophy that defines indigenous communities. This element of holism seems to have characterised the pre-industrial integrated curriculum, which was aimed at social empowerment and sustainability in the European context:

> The separation in the curriculum of mind from matter was the antithesis of village education in pre-industrial Europe. For many villages, the most highly prized individual was the chief artisan, such as the blacksmith, the carpenter, or the stone mason. Not only were they skilled in their craft but they also relied on them for practical community guidance in the social sense. The prowess of the artisan was deeply embedded in a social context that directly related to the natural environment from which his/her raw materials were derived. The artisan's prowess was necessarily defined by interdependent relationships found in the social, technical and environmental context of the craft. (Seemann 2000, p. 3)

In addition, "The context of human settlements generally dictated the things one had to know and become skilled in, in order to simply live" (Seemann 2000, p. 2). I thus hold a view that a holistic stance to education should be revisited because it offers a recipe for the integration of IKS and WKS (for more, refer to the three case studies cited earlier from Marchand 2008). The nineteenth century industrial revolution opposed to the integrationist approach, creating social strata based on status and oppression emulated by formal schooling system. For example, referencing the Indian context, Seemann writes that Ghandi once criticised the imposition of British education as a major contributor to the demise of rural India as a dynamic region of small cottage industry. Local innovation and small-scale rural productivity not only declined, but became less valued socially.

It seems that western societies will do well to revisit the educational system of the pre-industrial revolution to identify factors that promoted family and communal engagements, and ensured social development and sustainability. The London case study by Marchand (2008) attests to this possibility. As stated above, this provides a recipe for collaboration between indigenous and western communities.

## Arguments for Integrating Indigenous Technology in the Technology Education Curriculum

There are multiple reasons for integrating indigenous technology in Technology Education, and I present them in this section alongside an understanding of the possible interplay between indigenous knowledge and indigenous technology.

First, indigenous technology integration will help dispel misconceptions held about indigenous populations. Indigenous populations have a history of being referred to by the colonial masters as primitive, lower order, backward, ethnic minorities, marginalised, etc. (Odora Hoppers 1998). The Technology Education curriculum, through its collaborative and design-based approach, has a role to play in dispelling these misconceptions. This is possible through engaging both indigenous and non-indigenous learners in collaborative design projects facilitated through relevant pedagogy. The integration can also deal with the disjuncture between school and home evident in terms of less-than-satisfactory relationships between the world of the school and the child's world (home and community) expressed through terms such as 'gap', 'polarity', 'contradiction', 'distance', and 'discontinuity'—with resultant alienation from one's parents' community as a consequence of schooling (Sarangapani 2003). Technology Education therefore has potential to be a subject of hope, bridging the gap between home and school by addressing authentic problems in students' environments. In fact, the technology practiced in the local context should inform teaching.

Another reason for integrating indigenous knowledge in Technology Education is to confront the process of colonisation, with its system of aggrandisement. Colonisation outlawed or suppressed IKS, contributing significantly to the low levels of educational attainment of indigenous populations and high rates of social issues, such as suicide, incarceration, unemployment and family or community separation (Association of Canadian Deans of Education [ACDE] 2010, p. 2). Indigenous forms of technology can instil interest in indigenous learners as they will learn about their communities' contributions. Integration can help restore identity and culture. Further, traditional ecological knowledge is increasingly becoming highly valued by scientists and environmentalists—yet it is being lost through loss of identity and links with the land, marginalisation by WKS, ownership structures being devalued and traditional ecological knowledge used by outsiders for economic gain, and so on (Michie 1999). A culturally-responsive Technology Education curriculum has potential to help facilitate respect for indigenous communities by their western counterparts and instil assertiveness of indigenous learners. There are also benefits for non-indigenous learners–indigenous knowledge can enhance their understanding of indigenous peoples, alternative ways of looking at the world and valuing traditional ecological knowledge (Michie 1999). The integration of indigenous technology can also expand their knowledge and appreciation of other forms of technology.

The next reason for integration has to do with contemporary school systems and the prevalence of de-contextualised teaching. Western-dominated educational programmes can cause indigenous learners to reject or forget their cultural knowledge; many students attend boarding schools away from their villages and thus do not complete their education in their cultural context. The results are that indigenous students either end up being absorbed in conventional professional careers that detach them from their home, or they do not enter these careers, finding themselves lost somewhere between the traditional culture of their villages and the new culture of development (Michie 1999). Re-contextualised Technology teaching can produce professionals who will carry the onus to identify and actively contribute towards the development of their communities' technology.

To do this, teachers will need to adopt culturally-sensitive approaches to teaching that do not close down indigenous perspectives (Semali and Kincheloe 1999; Wlodkowski and Ginsberg 1995). Technology teachers should critically relate the subject matter and their teaching to the cultural milieu of all learners, including those with indigenous heritage. This calls for the construction of just and inclusive academic environments. Teachers should approach their teaching as hermeneuts (helping learners and other individuals to make sense of the world around them) and epistemologists (seeking to expose how accepted knowledge came to be validated) (Semali and Kincheloe 1999). This inclusive attitude will help transform research in the field of Technology Education by sensitising scholars towards the field of indigenous technology. It will also transform Technology teachers from being mere knowledge imparters and dispensers to becoming co-researchers with learners, and it will involve indigenous communities. As epistemologists, teachers will focus attention on ways knowledge is produced and legitimated. Creative ways to critically engage other forms of knowledge will ensure accommodation of indigenous technological perspectives. The project- and design-based approach of Technology Education offers excellent opportunities for students to explore, from different contexts and cultures, existing forms of technology, as well as future possibilities. Indigenous and non-indigenous learners can input into each other's contributions as they work on their design projects.

A further reason for integrating indigenous technology into Technology Education is to address concerns about the widening gap between rich and poor in the knowledge-based global economy (Carvallo 2000). The looting of indigenous knowledge and technology has contributed to this widening gap: Some pharmaceutical companies use indigenous knowledge to identify medicinal plants and extract the active ingredients and exploit them commercially, with little or no return to the owners of the knowledge; others exploit the genetic resources of plants cultivated by indigenous peoples for genetic materials that they have then patented (Michie 1999; Shizha 2006). This has violated indigenous people's intellectual property rights (Marinova et al. 2010). With its aim to produce critical and responsible learners, Technology Education can—and should—instil a character of respect for intellectual property and ownership of knowledge. Learners, through a Technology Education incorporating IKS, can learn about indigenous technology and begin to realise its value and the role it can play in the knowledge economy and global context. They can all learn how different forms of technology can add value to sustaining an economy that all can benefit from.

It can thus be deduced that indigenous technology has multiple roles to play in Technology Education curriculum and teaching—to address misconceptions that exist about indigenous people, relate teaching and learning to home and community knowledge, restore and affirm the identity and culture of indigenous learners, facilitate collaboration between indigenous and non-indigenous learners using relevant culturally sensitive pedagogies, and acknowledge the contribution that indigenous technology can make to the economy for all to benefit from.

How, then, might Technology Education curriculum and teaching integrate indigenous technology?

## A Technology Education Curriculum that Integrates Indigenous Forms of Technology

My take is that Technology teachers should integrate indigenous technology in the curriculum, with sufficient emphasis on indigenous technology to restore its lost status and value. This means that Ministries of Education, and teachers, have to re-consider the Technology Education curriculum that they offer to learners.

Let me borrow Seemann's (2000) technacy concept to suggest a Technology Education curriculum that integrates indigenous technology. This concept represents a paradigm shift from an industry-driven approach to Technology Education curriculum, which by and large reinforces socio-economic strata that elevate western elites at the expense of the poor, the majority of whom happen to be indigenous. Rather, it is about a holistic approach to Technology Education in order to promote social empowerment, development and sustainability. In other words, 'technacy' is a holistic approach to perceiving, teaching, practicing and learning technology; it is a holistic technology problem solving, communication and practice; it is a view that recognises technology as value laden; it is about an integrated approach to subjects—resembling the philosophy of life out there; and it is based on Dewey and Archambault's (1974) opposition to divisions of curriculum and claim that disintegrated school curricula produce disintegrated minds. In other words, curriculum should be designed in such a way that it embraces the realities of indigenous communities which tend to integrate subjects. This will help not to overemphasise the compartmentalisation of subjects in discrete forms. (See Chap. 10, this volume, for Cathy Buntting and Alister Jones' consideration of how technology can be aligned with other school subjects.)

Seemann (2000) argues that Design & Technology should be based on a framework that is socially innovative, and that it should maintain a link between learning and its application. Ironically, the theoretical model underpinning technacy emanates from western societies (it is based on the ideas of Hegel, Feuerback, Max, Dewey, Wortofsky, Schumacher, Papanak and Ihde). However, the model is attuned to the social learning styles and knowledge frameworks of indigenous Australians (Walker & Seemann in Seemann 2000). It also applies, in general, to other indigenous populations.

Based on the technacy concept and an integration of IKS and WKS, Technology Education might include the following:

- Central goal: To develop a learner who is a skilled, holistic thinker and doer who can select, evaluate, transform and use appropriate technologies that are responsive to local contexts and human needs (Seemann 2000). Such learners will have to be developed to adopt an open-mind approach to accommodate other forms of knowledge whilst ensuring respect for the philosophical profundities set in the local communities and those existent in other indigenous contexts.
- Content knowledge: Technological knowledge is packaged in such a way that it includes technologies existent in indigenous contexts sustaining the lives of people. The following pointers can help with the formulation of strands to be considered (amongst others):
  - the concept of technology as it relates to multiple contexts;
  - epistemological issues surrounding the concept of technology;
  - end-users' cultural values versus designers' cultural values;
  - technologies and designs that include indigenous contexts, including case studies;
  - technological resources and materials that include those in indigenous contexts;
  - principles of technological applications that embrace those in indigenous contexts;
  - profiles of prominent innovators and technologists, including those from indigenous contexts; and
  - trade value of the technologies, including those from indigenous contexts.
- Learning support materials and equipment: Design and put in place materials and equipment suitable to teach about both western and indigenous technology:
  - textbooks and other resources that represent the learners' technological milieu;
  - examples and learner activities that are attuned not only to western technology, but also to indigenous technology, and designers and innovators who learners can identify with;
  - incorporation of textbooks and other resources designed and written by indigenes who have a deep understanding of indigenous technology;
  - equipping Technology Education classrooms/labs with indigenous designers' and manufacturers' products; and
  - techno-labs designed in such a way that they reflect both western and indigenous technological worlds.
- Pedagogical approaches: A number of pedagogical approaches can be considered:
  - orientate teaching around culturally responsive pedagogy so as to be invitational to all learners;
  - teach about indigenous technology as packaged in the curriculum and relate it to regional and global contexts where appropriate;

- adopt a community model of co-teaching—invite community-based technologists (para-teachers or elders) to teach about and demonstrate the technologies that they employ in their own settings;
- plan learning activities so that learners will engage actively with community members (especially information-rich elders) in design projects—this process will ensure principles of ubuntu and community of practice as well as social cohesion, development and sustainability;
- adopt flexible approaches to design, considering tacit knowledge existent in elders. Some designs can be done through mentoring and sharing of expertise by elders. Thus, the idea of design extended to a grassroots level enters because of emergent design (Carvallo 2000)—design of the learner's own interest which is relevant and applicable in his/her own context guided by the wisdom of elders; and
- critical discourses on epistemological issues surrounding technology.

• Assessment: Plan assessment in such a way that it engages learners in content that they can identify with. Assessment should also be related to the learning materials and content above. It should target applied knowledge, be context sensitive and help to graduate the learner as an expert.

## Challenges to Teachers

At this stage I want to reflect on a few challenges that teachers may face in their efforts to integrate indigenous technologies in the curriculum and teaching. The first has to do with attitude. Teachers should begin to show interest in other forms of knowledge and technology in order to accommodate all learners in their teaching, irrespective of the cultural background. Second, it should be the aim of every teacher to respect the knowledge that each learner brings to class, and to tap into it to enrich the teaching and learning activities. Third, non-indigenous teachers should want to treat indigenous teachers as mentors and producers of pedagogical knowledge. Finally, teachers should take advantage of building relationships within and beyond the school—interacting with communities and elders to exchange knowledge and wisdom.

## Conclusion

The reality is that the technological world comprises both modern and indigenous forms of technologies. Thus, the future of Technology Education should be shaped by the integration of indigenous technology. Technology learners stand to benefit tremendously by being exposed to both worlds. For example, consider the *Zulu*

**Fig. 4.1** The iconic Zulu Mama Chair is an integration of South Africa's first and third world reality by combining indigenous Zulu basket weaving craft with modern materials. The basket seat expresses the archetypal feminine activity of gathering, an appropriate gesture for indoor and outdoor café seating. The weaving work contributes to the economic empowerment of township crafters. The frames are made from rustproof, 60 % recycled stainless steel and can be finished in a variety of powder coated colours or polished stainless steel. The UV stable polyethylene plastic weaving material is also available in various colours. The *black coloured* plastic is made from recycled factory waste (http://haldanemartin.co.za/zulu-mama-chair-2/)

*Mama Chair* example below, which provides an opportunity for learners to learn from both IKS and WKS (Fig. 4.1).

In order for all learners to learn about indigenous technologies in an authentic and honourable way, I have argued in this chapter for the integration of indigenous technology into Technology Education. To do this, I have offered reasons for integrating indigenous technology, dichotomising between WKS and IKS for the sake of understanding the differences between the two for purposes of integration. Most importantly, I have presented a scenario for an integrated approach moving forward. In order for this to happen, teachers need to value the role that indigenous technology plays and integrate it in their teaching. This will arouse interest in learners, especially indigenous learners, many of whom are disenfranchised by lessons that do not integrate a representation of their indigenous world.

Technology is closely tied to the context in which it plays out, and this presents a golden opportunity to Technology teachers to integrate varied forms of technology, in this case IKS-based technology and WKS-based technology. There are multiple practical examples of indigenous technology in indigenous contexts, and teachers could harness this by designing projects for their learners that are related to these contexts. In the process, students could consider design projects that integrate the wisdom that elders can impart, either by interacting with communities or by inviting these elders to demonstrate their technological profundity in the class.

The outcome could be students who are empowered, open to other forms of knowledge, can work comfortably on design projects as they exchange ideas from multiple perspectives, and can tolerate each other as they work as teams and appreciate each other's ideas manifested through projects designed from varied contexts.

## References


Adams, H. H., III. (1983). African observers of the universe: The Sirius question. In I. V. Sertima (Ed.), *Blacks in science: Ancient and modern* (pp. 27–46). New Brunswick: Transaction Books.

Apple, M. (2000). *Official knowledge* (2nd ed.). London: Routledge.

Asante, M., & Asante, K. (1983). Great Zimbabwe: An ancient African city-state. In I. V. Sertima (Ed.), *Blacks in science: Ancient and modern* (pp. 84–91). New Brunswick: Transaction Books.

Association of Canadian Deans of Education [ACDE]. (2010). *Accord on indigenous education.* http://educ.ubc.ca/sites/educ.ubc.ca/files/FoE%20document_ACDE_Accord_Indigenous_Education_01-12-10.pdf. Accessed 18 June 2013.

ATPS. (2010). *The African manifesto for science, technology and innovation.* Nairobi: African Technology Policy Studies.

Atte, O. D. (1992). *Indigenous local knowledge as a key to local level development: Possibilities, constraints, and planning issues* (Studies in technology and social change). Iowa: Iowa State University Research Foundation.

Barab, S. A., Barnett, M., & Squire, K. (2002). Developing an empirical account of a community of practice: Characterizing the essential tensions. *The Journal of the Learning Sciences, 11*(4), 489–542.

Battiste, M. (2002). *Indigenous knowledge and pedagogy in first nations education: A literature review with recommendations.* Ottawa: National Working Group on Education and the Minister of Indian Affairs. www.afn.ca/uploads/files/education/24._2002_oct_marie_battiste_indigenousknowledgeandpedagogy_lit_review_for_min_working_group.pdf. Accessed 1 Mar 2012.

Carvallo, D. (2000). Emergent design and learning environments: Building on indigenous knowledge. *IBM Systems Journal, 39*(3&4), 768–781.

Castiano, J. P. (2011). *Researching in African IKS context.* Paper presented at the African indigenous knowledge systems conference, Windhoek, Namibia.

Cheek, D. W. (1992). *Thinking constructively about science, technology and society education.* Albany: State University of New York.

Coelho, E. (1998). *Teaching and learning in multicultural schools.* Clevedon: Multilingual Matters.

Custer, R. L. (1995). Examining the dimensions of technology. *International Journal of Technology and Design Education, 5*(3), 219–244.

Dewey, J., & Archambault, R. D. (1974). *John Dewey on education: Selected writings.* Chicago/London: University of Chicago.

Emeagwali, G. (2003). African indigenous knowledge systems (AIK): Implications for the curriculum. In T. Falola (Ed.), *Ghana in Africa and the world: Essays in honour of Adu Boahen* (pp. 121–137). Trenton: Africa World Press.

Gegeo, D. W., & Watson-Gegeo, K. A. (2002). Whose knowledge? Epistemological collisions in Solomon Islands community development. *The Contemporary Pacific, 14*(2), 377–409.

Green, L. J. F. (2008). 'Indigenous knowledge' and 'science': Reframing the debate on knowledge diversity. *Archaeologies: Journal of the World Archaeological Congress, 4*(1), 144–163.

Gumbo, M. T., & Williams, P. J. (2012). Discovering South African teachers' pedagogical content knowledge. In *20th annual meeting of the Southern African Association for Research in Mathematics, Science and Technology Education*, Lilongwe, Malawi.
Hall, M. (1996). *Archaeology Africa*. Pretoria: David Phillip.
Jonathan, L. T. (1996). Traditional versus modern medicine: The case for a collaborative approach to primary health care (PHC). *Journal of Research: Ethnomedicine in Southern Africa, 4*, 9–18.
Kimbell, R. (2008). Indigenous knowledge, know-how, and Design & Technology. *Design and Technology Education: An International Journal, 10*(3), 8–10.
Lawton, D. (1982). Knowledge and curriculum planning. In T. Horton & P. Raggatt (Eds.), *Challenge and change in the curriculum* (pp. 30–36). Kent: Hodder and Stoughton.
Maluleka, K., Wilkinson, A., & Gumbo, M. (2006). The relevance of indigenous technology in Curriculum 2005/RNCS with special reference to the Technology Learning Area. *South African Journal of Education, 26*(4), 501–513.
Marchand, T. H. J. (2008). Muscles, morals and mind: Craft apprenticeship and the formation of person. *British Journal of Educational Studies, 56*(3), 245–271.
Marinova, D., Lozeva, S., & Seemann, K. (2010). Community conversations for sustainability in the desert: Leonora, Western Australia. *Journal of Economic and Social Policy, 13*(2), 103–122.
McCade, J. M., & Weymer, R. A. (1996). Defining the field of technology education. *The Technology Teacher, 55*(8), 40–46.
Michie, M. (1999). *Where are Indigenous peoples and their knowledge in the reforming of learning, curriculum and pedagogy?* Paper presented at the 5th UNESCO-ACEID international conference, Bangkok, Thailand.
Mokgoro, J. Y. (1997). Ubuntu and the law in South Africa. *PER/PELJ, 1*(1), 14–26.
Msila, V. (2009). Africanization of education and the search for relevance and context. *Educational Research and Review, 4*(6), 310–315.
Mugabe, J. (n.d.). *Intellectual property protection and traditional knowledge: An exploration in international policy discourse.* http://www.oapi.wipo.net/export/sites/www/tk/en/hr/paneldiscussion/papers/pdf/mugabe.pdf. Accessed 3 Feb 2012.
Nakpodia, E. D. (2010). Culture and curriculum development in Nigerian schools. *African Journal of History and Culture, 2*(1), 90–98.
Obikeze, D. S. (2011). *Indigenous knowledge systems and the transformation of the academy in Africa: The CULPIP Model.* http://citeseerx.ist.psu.edu/viewdoc/download?doi=10.1.1.164.9357&rep=rep1&type=pdf. Accessed 30 Nov 2013.
Odora Hoppers, C. A. (1998). *A comparative study of the development, integration and protection of Indigenous knowledge systems in the third world: Draft proposal and project plan.* Unpublished paper.
Odora Hoppers, C. A. (2002). Introduction. In C. A. Odora Hoppers (Ed.), *Indigenous knowledge and the integration of knowledge systems* (pp. vii–xiv). Claremont: New Africa Books.
Omolewa, M. (2007). Traditional African modes of education: Their relevance in the modern world. *International Review of Education, 53*, 593–612.
Orevbu, A. O. (1997). *Culture and technology: A study on the 1997 theme*. Paris: UNESCO.
Owuor, J. A. (2007). Integrating African indigenous knowledge in Kenya's formal education system: The potential for sustainable development. *Journal of Contemporary Issues in Education, 2*(2), 21–37.
Potgieter, C. (1998). *Technology Education: Enhancing the best of both worlds.* Paper presented at international Best-of-Both-Worlds conference, Pretoria, South Africa.
Rains, F. V. (1999). Indigenous knowledge, historical amnesia and intellectual authority: Deconstructing hegemony and the social and political implications of the curricular "other". In L. M. Semali & J. L. Kincheloe (Eds.), *What is indigenous knowledge: Voices from the academy* (pp. 317–331). New York: Falmer.
Robyn, L. (2002). Indigenous knowledge and technology: Creating environmental justice in the twenty-first century. *The American Indian Quarterly, 26*(2), 198–220.

Rowlands, M., & Warnier, J. (1996). Magical iron technology in the Cameroon grassfields. In M. J. Arnoldi, C. M. Geary, & K. L. Hardin (Eds.), *African material culture* (pp. 51–72). Bloomington: Indiana University.

Sarangapani, P. M. (2003). Indigenising curriculum: Questions posed by Baiga Vidya. *Comparative Education, 39*(2), 199–209.

Seemann, K. W. (2000). *Technacy education: Towards holistic pedagogy and epistemology in general and indigenous/crosscultural technology education*. Paper presented at the international technology education research conference, Gold Cost, Australia.

Semali, L. M., & Kincheloe, J. L. (1999). What is indigenous knowledge and why should we study it? In L. M. Semali & J. L. Kincheloe (Eds.), *What is indigenous knowledge? Voices from the academy* (pp. 3–57). New York: Falmer.

Sertima, I. V. (1983). The lost sciences of Africa: An overview. In I. V. Sertima (Ed.), *Blacks in science: Ancient and modern* (pp. 7–26). New Brunswick: Transaction Books.

Shizha, E. (2006). Legitimizing indigenous knowledge in Zimbabwe: A theoretical analysis of postcolonial school knowledge and its colonial legacy. *Journal of Contemporary Issues in Education, 1*(1), 20–35.

Tedla, E. (1995). *Sankofa: African thought and education*. New York: Peter Lang.

Vandeleur, S. (2010). *Indigenous technology and culture in the Technology Education curriculum: Starting the conversation—A case study*. Unpublished doctoral thesis, Rhodes University, Rhodes, South Africa.

Van Wyk, J. (2002). Indigenous knowledge systems: Implications for natural science and technology teaching and learning. *South African Journal of Education, 22*(4), 305–312.

Williams, P. J., & Gumbo, M. T. (2011). Discovering New Zealand technology teacher's PCK. *Journal of Technology Studies, XXXVII*(1), 50–60.

Wlodkowski, R. J., & Ginsberg, M. B. (1995). A framework for culturally responsive teaching. *Educational Leadership, 53*(1), 17–21.

Zulu, I. M. (2006). Critical indigenous African education and knowledge. *The Journal of Pan African Studies, 1*(3), 32–49.

# Chapter 5
# The Pedagogical Ecology of Technology Education: An Agenda for Future Research and Development

**David Mioduser**

In order to explore pedagogy for future technology education, this chapter weighs up the need to consider what we already know about effective pedagogical models for technology education with what might be required of technology education in the future, and how these goals might be addressed. The social, economic and cultural realities of this century are undergoing such radical changes, including technological transformations, that a thorough revision of teaching and pedagogical approaches is necessitated. The agenda for the future should include a great deal of research and development into addressing questions related to individual, group and social diversity. To pursue this agenda, a research and development pedagogical ecology framework is proposed that includes conceptual, contextual, pedagogical resources, and planning and implementation dimensions.

## Introduction

Discussing teaching and pedagogical issues in the context of the future of technology education (TE) is a challenging task. On one hand, it could be argued that valuable pedagogical approaches and models have already been identified and defined over recent decades as a result of systematic research. In this case, the challenge in the near past has been, and will surely be in the near future, not theoretical or conceptual but practical: how to **implement** the already consolidated pedagogical models and solutions in educational systems worldwide. Or in different wording, why what has already been proven as appropriate in research contexts and limited scale implementations has not been widely adopted, and what is required for this to happen. On the other hand, a contrasting perspective suggests that the social, economic and

D. Mioduser (✉)
School of Education, Tel-Aviv University, Tel Aviv, Israel
e-mail: miodu@post.tau.ac.il



PJ. Williams et al. (eds.), *The Future of Technology Education*, Contemporary Issues in Technology Education, DOI 10.1007/978-981-287-170-1_5

cultural realities of this century are undergoing radical changes, including technological transformations, thus demanding a thorough revision of the goals and means of TE in general, and of teaching and pedagogical approaches in particular.

This chapter attempts to integrate the contrasting perspectives with the aim of exploring the main pedagogical issues and challenges to be faced in TE in the near future. It should be clear from the outset that it is difficult to discuss pedagogical issues in isolation from content-related and curricular issues within which pedagogies are developed and implemented. Many of these are explored in detail in other chapters in this book. I will try, however, to refer to these aspects only when required and as background for the elaboration on the main pedagogical issues.

## Pedagogical Issues and Challenges in Technology Education

Before elaborating on specific issues related to TE, let me depict briefly the general framework concerning learning and pedagogy within which the specific issues are discussed.

In this chapter I advocate for a meaningful integration of several theoretical standpoints in support of a view of **learning as a complex system**. This systemic phenomenon compromises factors and processes related to the individual's cognitive and affective states (e.g., existing knowledge, schemes and beliefs; developmental stage; cognitive and learning style; perception of environmental—physical and human—traits; motivational foci), as well as to the world's states (e.g., others; social and cultural—including material—landscape; events, processes, contingencies and demands; and affordances of the socio-cultural context). No single theoretical standpoint (e.g., individual-centred constructivism[1] or social constructivist approaches) can grasp the complexity of the intricate relationships among individual and social factors affecting learning and cognitive development. As Rogoff and Chavajay (1995) have suggested,

> . . . individual, social, and cultural levels are inseparable. Analysis may focus primarily on one but not without reference to the others as if they can exist in isolation [. . .] the aim is to understand the developmental processes involved in activities involving individual, interpersonal, and community/cultural processes. (pp. 872)

---

[1] Constructivist pedagogies are based on a theory of learning—constructivism—that claims that learning results from the active involvement of the learner in meaning-making and knowledge construction (rather than from passive transmission/reception of information). New knowledge is constructed by the learner upon her/his previous knowledge schemas and resources. Roughly, two broad approaches towards the knowledge construction process have emerged: psychological constructivism, focusing on the ways meaning is constructed in the individual's mind; and social constructivism, emphasising the role of social and cultural forces affecting the meaning construction process. Constructivist pedagogies aim to facilitate and support learners' active participation in knowledge construction by means of unique methods and learning environments (e.g., inquiry and design tasks, building kits, productivity or modelling software).

Along these lines, Perkins (1993) proposes a "person-plus" view (in contrast to a person-solo view) for understanding cognitive growth and learning processes in terms of the dynamic synergy between the individual traits and the human and physical surrounds. In addition, Salomon (1993) addressed the need to "overcome the situational determinism I see in the radical view of distributed cognitions and the intrapersonal determinism in the radical solo view of cognitions" (pp. xviii), adopting an integrative and interactive perspective about the co-evolvement of intra-personal and inter-[personal, social, cultural] developments.

Stemming from this theoretical stance, pedagogical solutions are developed following clearly defined principles:

- Learner-centred approach: the learner is not only the primary target of the pedagogical process, but also the main engine driving it
- Knowledge is gradually constructed by the learner as a result of meaningful interactions between existing knowledge and schemes and new triggering and challenging demands embedded in the pedagogical situation (O'Donnel 2012). The creation of perceivable challenging gaps between existing and required knowledge are regarded as powerful pedagogical means for supporting knowledge construction processes.
- Knowledge is also gradually constructed by the learner as a result of meaningful interactions with a rich and conceptually challenging environment (Richardson 2003). The context, characteristics and conditions of the environment in which learning takes place (e.g., social, cultural or geographical properties and materials) offer both affordances and challenges that can empower the knowledge construction process. Consequently, pedagogical solutions ought to be devised to afford the manipulation and use of plentiful resources, such as tools, materials or construction kits—whether physical or, in the new digital reality, virtual.
- The socio-cultural perspective: the view of knowledge as a social construct, grants peer interaction and collaboration a substantial role in supporting the collective creation of knowledge assets (Rogoff and Chavajay 1995). When the construction of knowledge is conceived as resulting from the interaction among peers holding different levels of expertise within a specific context (e.g., given social, cultural, economic or political realities, and even the quality of knowledge and resources available), pedagogical solutions are devised to support such collaborative work in continuous dialogue with the environmental context.
- An additional angle relating to the social aspects of learning is supplied by the constructionist approach (Papert and Harel 1991). From this standpoint, the creation of a 'public entity' (e.g., an artefact, a computer programme, a theory) means the externalisation and objectification of inner (to the mind) thinking processes and knowledge. In this process, the learner's externalised thoughts are open to peer discussion, critical evaluation, and insightful comments and suggestions, thus promoting both individual and group learning.

If the integrative approach is of value for understanding learning as a complex system, it is equally important for the development of pedagogies that address in holistic fashion all its different aspects—what is defined in this chapter as "pedagogical ecology". Such an ecology encompasses the range of human and physical resources and methods aiming to support learning in all of its individual and social dimensions.

Moving from this more general elaboration about learning and pedagogy to its instantiation in technology education, the following sections briefly introduce and explore some of the key issues that are relevant to the pedagogical ecology of TE.

## *To Learn* Technology Is to *Learn to Think* Technology

A long-standing conceptual stance has dominated—and still does—the rationale for TE: the socio-technological approach (see, for example, Dakers 2006; Petrina 2000). This rationale embraces TE in its various forms, including what has been defined as "technological literacy" or "technology for all" programmes as well as expertise-oriented programmes (e.g., high school and higher education specialisation studies). The main drivers of this rationale call on social and economic considerations: Technology (and often science and technology) is an essential asset in evolving twenty-first century economies; technologically knowledgeable and skilled citizens contribute to a country's economic competitiveness; technology-related knowledge and skills are important resources for individuals' successful integration in contemporary workplaces and for ensuring their social mobility; and technology itself drives technological development and progress.

Following the above rationale, a range of pedagogical solutions has been developed. In one model, oriented towards what is conceived as needs of the economy and the workplace, technology teaching emphasises the development of craft-type skills and the acquisition of procedural knowledge rather than conceptual knowledge and the thinking behind the technological processes. Typical examples are lesson plans fostering the reproduction of teacher- or expert-made construction plans and design solutions, or tasks involving the manipulation of pre-set collections of materials or building-pieces. On occasion, background and contextual information relevant to the tasks are supplied as informative texts (rather than challenging the student with the need to gather, analyse and synthesise information), often in the form of "about technology" boxes in science textbooks. The 'pedagogical spirit' behind this view of TE as a means for educating future workers or actors in technology-based economies emphasises products over processes and measurable (assessable) end results over knowledge-construction processes (Brophy et al. 2008; Fritz 1996).

A contrasting pedagogical perspective moves away from the socio/technological approach towards a cognitive/epistemological approach (Mioduser 2009). The philosophical stance behind this approach relies on the view of technology as a defining characteristic of human beings' thinking and intellectual development (Nye 2006; Preiss and Sternberg 2005). Technology, here, refers not only to the sets of skills and knowledge as defined in the technological disciplines—the formal body of technological knowledge. It also refers to humankinds' stance (intellectual as well as affective) towards the creation of the made world out of the natural world in response to needs and opportunities, and to the thinking processes and capabilities involved in these acts of creation. From this perspective, therefore, learning technology involves learning about thinking, learning, the (outer) invented world vis-à-vis the (inner) inventor's world, and the results of the co-evolution—both phylogenetic and ontogenetic—of the designed and the designer's worlds. As Cole and Derry (2005) conclusively claimed: "We have met technology and it is us" (pp. 209).

Consequently, pedagogical solutions underpinned by a cognitive/epistemological approach emphasise processes over products. Lesson plans and tasks are devised as challenging situations demanding consecutive cycles of action and reflection. Materials supplied take the form of problem-solving playgrounds rather than pre-set building plans. Inquiry, exploration, analysis and synthesis cycles, and personal involvement in the very planning of the path leading from the existing situation to the desired situation (reaching the solution) are key components of this pedagogical form. Often, the process involves collaborative work, peer critical/reflective discussions, peer and teacher formative feedback, and whole-process assessment. In such a pedagogical programme, the body of target skills and knowledge exceeds the strict boundaries of formal disciplinary knowledge to encompass higher-order and metacognitive sets of skills such as these involved in system thinking, problem solving, and holistic perceptions of technological problems and solutions (Walmsley 2003; see also Chap. 8, this volume, by David Barlex and Chap. 9 by David Spendlove).

It is obvious from the above that a significant pedagogical challenge for technology education is to develop a balanced approach in which learning of relevant skills and information (i.e., formal technological knowledge and skills) is supported, while at the same time emphasising knowledge-construction processes by the learners themselves. I return to the discussion of these pedagogical questions in a later section of this chapter.

## *To Learn* Technology Is *to Do* Technology

As obvious as this claim—To learn technology is to do technology—may seem, there is a significant gap between it and the pedagogical reality in many TE classrooms worldwide (Sherman et al. 2010).

The obstacles for implementing pedagogical solutions centred in *doing* technology are diverse in nature. In some cases there are logistical constraints that discourage teachers from planning classes centred in doing and making, for example, lack of time in the teaching schedule, lack of appropriate infrastructure for carrying out construction tasks, or—even more worrying—lack of appropriate background and training, particularly among teachers at the elementary and intermediate levels.

In other cases, where technological topics appear within integrative units (e.g., science and technology or social sciences), teachers often adopt a pedagogical approach in which technology-related content is delivered mainly in expository mode. In this approach, the dominant pedagogical resources are texts and visuals (e.g., from textbooks or the Internet), stories of and about technological developments and salient characters (e.g., inventors, entrepreneurs), and demonstrations (e.g., by the teacher in class, in museums, or using video clips retrieved from the Internet). Technology, in such contexts, is perceived as an 'informative' curricular subject and most activities rely on the manipulation of texts and visual materials, class discussions and, consequently, assessment of declarative knowledge gains.

In spite of these realities, there is ample consensus among practitioners and researchers in the TE community that students' active engagement in doing technology is an essential requirement for meaningful learning. This consensus is rooted in the theoretical standpoints briefly surveyed above, which emphasise the integration of constructivist as well as sociocultural theoretical perspectives about teaching and learning processes as complex systems (John-Steiner and Mahn 1996; Prawat 1996).

In addition, strong support for 'doing' technology stems from another theoretical standpoint focusing on the essence of technological development processes. Technology is essentially a thinking-and-making process in which action and reflection are intertwined and the resulting knowledge takes the form of both conceptual knowledge and its instantiation in an artefact, whether concrete or abstract, physical or symbolic. It is natural, therefore, to claim that the most logical way to be engaged in learning technology is to be engaged in *doing* technology—in experiencing the authentic process of creating a technological solution in response to a need or opportunity. Here, the pedagogical challenge lies in creating authentic learning contexts that can be holistically experienced by learners as credible and immersive and during which they create their own conceptual as well as artefact-embedded technological knowledge. In such pedagogical approaches the key idea of 'the creation of the made world' is conveyed through situations that allow learners to themselves be involved in manipulating and creating aspects of the made world.

## *Many "Technology Educations"*

An elaboration on pedagogical issues and challenges in TE should take cognisance of the varying goals pursued for different ages, populations and target levels of expertise. For students in the school cycle (i.e., kindergarten to high school, and in some countries beyond this into tertiary pre-college institutions), roughly four emphases can be found: technological literacy, vocational preparation, expert specialisation and professional studies. Although some pedagogical principles are relevant to all these emphases (e.g., a learner-centred approach, affording student involvement in doing and making processes), others are unique to a specific curriculum emphasis.

For example, regarding technological literacy, pedagogical solutions are designed considering age and developmental aspects (literacy is usually targeted to the younger population), addressing the development of conceptual understanding and a technological worldview rather than formal expertise in a technological field. Usually these programmes focus on aspects of the made world at the conceptual and phenomenological levels (rather than at formal-disciplinary levels); on the conduction of simplified design processes (in the form of simple didactic versions of complex engineering-design processes); and on the elaboration of ethical, moral or social issues stemming from technological developments. Correspondingly, pedagogical practices include activities that aim to direct learners' awareness to technological phenomena and the multiple issues related to these, and design tasks that conceptually resemble processes in the technological world but are conducted within the constraints of the classroom (e.g., concerning available—and safe—infrastructure, materials, tools). 'Doing' tasks, even if highly active and fostering authentic immersion in hands-on processes, are to a large degree open-ended and provide learners with opportunities to explore diverse paths in developing their conceptual understanding of processes.

In contrast, pedagogies implemented in specialisation or trade-acquisition programmes (e.g., in vocational and practical-engineering programmes) are far more formally structured and explicitly oriented towards defined sets of skills and bodies of expert knowledge. This is not to imply that, in contrast to literacy-oriented pedagogy, expertise-oriented pedagogy should necessarily fall in the realm of 'traditional' and product-only oriented practices (as appears to be the reality in most expert-formation institutes worldwide, see Sjoberg 2001; Walmsley 2003). An illustrative, still challenging example is the almost century-old pedagogical approach implemented in the German Bauhaus school (Gropius 1919; Itten 1975).

Although the Bauhaus functioned for a relatively short period (1919–1933), its impact on design, architecture and various arts has been of crucial significance—not least because of the pedagogical philosophy and teaching practices developed by its teachers (themselves recognised artists, designers and architects). The school's curriculum, depicted as a series of concentric circles (an iconic entity representing the school's curricular philosophy), presents in the outer circle the basic or preliminary course, then the series of workshops focusing on diverse materials and technologies, in to the holistic integration of all components in the most inner circle: the artefact. The learning journey therefore integrated three main knowledge components: crafts, representational means (drawing, painting), and theoretical and scientific principles in the natural sciences and technology. Tasks and assignments aimed to develop the students' minds as well as hands, and their gradual mastery of concepts as well as skills. An additional important layer accompanying the formation process related to the social and historical context within which the work (a design, a building, a piece of art) is performed. The product of the formation process—the craftsperson, the designer or artist—should feel and be integrated in the community, and be aware of the social, economic, ethical and historical implications of their work. Correspondingly, the tasks faced were authentic in character and demanded attention to a whole spectrum of aspects—from the purely creative, through all facets of the design process, to industrial and commercial issues. In other words, the Bauhaus succeeded in fostering the development of professional designers and artists through a high quality and complex "constructivist" (in contemporary terms) pedagogical system, which has had significant subsequent impact on the way the holistic development of professional designers is conceived and planned.

Drawing on a completely different context, the “implicit pedagogies” that characterise the way novices gradually acquire mastery of professional skills in the workplace and informal settings are also of interest. These pedagogies have been linked to theoretical approaches emphasising “situated cognition” and “apprenticeship” processes (Collins 2006; Seely Brown et al. 1989). The main claim is that

> . . . for most of history . . . children learned how to speak, grow crops, construct furniture, and make clothes. But they didn’t go to school to learn these things; instead, adults in their family and their communities showed them how, and helped them to do it. (Collins 2006, p. 47)

Of course, apprenticeship as a way of acquiring skills and expertise is still relevant not only in many societies worldwide (concerning the traditional apprenticeship involved in learning a trade or a craft), but also in the context of the most advanced technologies. For instance, a great deal of knowledge and a large array of capabilities are acquired by children while interacting with and manipulating technologies in authentic real-world tasks and in informal settings, with the support of peers and adults. Similarly, many training and knowledge-upgrading processes are carried out in the workplace when facing real situations (i.e., the tools, methods, problems) and interacting with relevant experts. A challenging question for TE is how the features of these “real-world-pedagogies” can be harnessed to design classroom-based pedagogies with the aim of bridging between formal and informal learning processes.

## *Are There “Obvious” Pedagogical Models in Technology Education?*

Without any doubt, much of the content included in any technology curriculum is taught using pedagogical methods commonplace across school subjects, and large segments of declarative and procedural knowledge are conveyed through tasks, exercises and text-based assignments similar in character to the majority of other learning activities in school. However, given the nature of technology education, there are several pedagogical models that appear to fit more naturally with the learning purposes.

These pedagogies stress aspects such as: engagement in doing; a systems perspective; emphasis on processes; non-linear paths (both for accomplishing tasks and for learning); collaborative work; and rich information-based decisions and actions. Well known examples of such pedagogical models extensively discussed in the educational research literature include ‘Learning-by-design’, ‘problem- and project-based learning’, and ‘case-based learning’ (Kolodner et al. 2003; Williams and Williams 1997).

Proven pedagogical models that have been implemented in large-scale programmes in different countries include the Nuffield project (see Barlex 2008) and the Israeli ‘New MABAT’ curricular project (Nachmias et al. 2007–2010). The

Nuffield approach follows a powerful motto: "Capable pupils can design what they are going to make and then make what they have designed" (Barlex 1998, pp. 143). The building blocks of the curriculum are tasks of two main types: resource tasks and capability tasks. Resource tasks ("small tasks") focus on knowledge and skills required in designing and making assignments; capability tasks ("big tasks") are the designing and making assignments. Effective learning demands learners' involvement in a sequence of making and designing tasks (rather than a single experience) characterised by a "mixed diet" of small and big tasks. Intertwined in the sequence are "case studies"—true stories about design and technology in the world outside the school—allowing learners to understand how products are designed, manufactured and marketed, and that they impact on the lives of individuals and societies.

Common to the vast majority of TE pedagogical models is a scenario in which the learner faces a problematic or challenging situation, or a need to be satisfied, and becomes engaged in a multifaceted process leading to the solution of the problem or the creation of a product—also known as the 'design process'. A key issue that has been a matter of debate among researchers concerns the structure and rigidity of the process to be followed by the learners. In many textbooks and proposals for lesson plans, the design process (or in some cases the project completion process) is depicted as a structured and almost linear sequence of stages leading from the initial step (usually "identification and definition of the problem") to the final step (usually "evaluation of the result") (Mawson 2003).

Contrasting views in the literature raise questions as to the appropriateness of the structured model, both as a valid representation of real-world processes and as a pedagogical method (e.g., Mawson 2003; Mioduser and Dagan 2007). The main claim is that neither real-world designers and engineers, nor learners, proceed in strictly structured and linear paths when solving a problem or designing a solution. On the contrary, they follow complex, iterative and cyclical paths, moving among stages, functions and methods as required by the evolving solution-creation process. For example, evaluation and critical analysis, rather than being a "stage" at the end of the cycle, are functions continuously activated at different stages along the working process.

In contrast to the structured approach, a 'functional' approach has been proposed as more closely representing real-life design and technological problem solving (Mioduser and Dagan 2007). In this approach the different design and problem solving methods and resources comprise a 'toolkit' from which learners choose the appropriate 'tools' according to their needs. Research indicates that in the structural-linear approach, learners concentrate on a given stage at a time, often losing the view of the whole process or missing the contribution of the given stage to the solution. In contrast, the functional approach is more cognitively demanding, transferring the responsibility of constructing the solution path to the learners and requiring them to maintain a holistic mental picture of the various resources and stages.

No matter which approach is adopted, there is ample consensus about the pedagogical value of learners' active engagement in design and problem-solving processes. There is also increasing consensus as to the need to define the pedagogical goals of the tasks in pursuit of the mastery of cognitive as well as metacognitive skills, and fostering a systems view of the solution pathway. More often than not, design tasks are ill-defined and the complex process required for their successful completion cannot be reduced to a simple linear path nor to a simple set of skilled actions. Design-, project-, problem- and case-based assignments that make possible the acquisition and activation of a wide range of skills (e.g., planning, crafting, tools and materials manipulation, information gathering, adaptation and use) have therefore become powerful pedagogical resources in many technology education curricula.

## *Coping with Diversity vs. Celebrating Diversity*

In our postmodern times, during which awareness of the multiple social, cultural and political currents coexisting and flowing within and among societies has intensified, it is important that educators address substantive questions related to diversity (Darling-Hammond 2007; Lubienski 2003). Roughly speaking, two contrasting approaches to addressing these questions can be identified: *coping with* versus *celebrating* diversity. The first approach typically leads to the search for methods and solutions to what is perceived as problematic, for example, the performance of low achievers in technology education. The second approach represents a need to support individual growth on the basis of each person's strengths (while also being aware of each person's constraints). Such approaches can also be identified when considering diversity among societies (and countries) of varied socio-economic, cultural and political realities. In this context, educational and pedagogical questions are related not only to the choice between approaches but also, even within the 'celebrating diversity' approach, to how to ensure a balanced perception of the interplay between strengths and limitations in order to design appropriate pedagogies to support individual students' learning.

Of course, 'diversity' when discussing pedagogical approaches (and an agenda for the future of these in TE) can take various forms. First, there is diversity among students. Second is diversity among groups in the same society or country. Third, and from an even broader view, is diversity among cultures and among societies and countries. Each of these dimensions pose serious questions to educators and policy makers in educational systems. TE will not escape the need to face these questions as well.

At the individual and group levels two main standpoints are briefly identified, each of which differently affects the design of pedagogical methods and resources. The first, of frequent use by educational policy-makers and researchers, is the *statistical* approach. Within this approach, large populations are assessed using standardised instruments deemed to be highly reliable. As a result, individuals are grouped and placed in scales of varied kinds (e.g., achievement in technology) and correlational analyses are conducted to identify covariance among substantial variables (e.g., achievement and socio-economic status). The immediate result, for our purpose, is the identification of diversity parameters characterising groups of students, and the development of pedagogical solutions based on this identification, for example, remedial programmes or 'basic level' curricula for the low-achievers and (perhaps less commonly) enrichment tracks for the high-achievers.

One critical observation of this approach is that when the individual becomes a 'statistical individual' placed in a scale, her or his prospects determined by correlational data, then we lose the real learner with her or his personal deviation from statistical means, and her or his complex grid of strengths and weaknesses. For the system, the actual learner is no longer the basis for pedagogical decision making (e.g., as happens in tracking procedures in many countries). A second critical observation is that what to measure as the basis for systemic decisions is an issue that is socially and politically fraught. It is no secret that strong emphasis is placed on subjects such as mathematics, science and language (and moreover, the English language) as the leading subjects to be supported in educational systems in response to the perceived demands of the world's economic reality. This trend is supported by the "race" in which many countries engage by participating in international comparative studies of achievements in these subjects. As a result, the scales within which learners are placed are unfairly unidimensional, focusing mainly on what is conceived as 'leading subjects'. In other words, a complex and multidimensional creature, the learner, is assessed, classed and assigned pedagogical solutions on the basis of a narrow assessment of her capabilities.

The contrasting approaches towards diversity among individuals can be depicted metaphorically using the image of sliders. In the statistical approach, there is one slider within which learners (or groups of learners) are placed according to their performance (achievement), most probably resulting in a normal distribution along the scales values. As an alternative, perhaps each learner could be conceived of as a collection of sliders, as in a sound equaliser console, each slider representing a specific capability or trait. Some will indicate high values, others lower values—each individual learner is the configuration of values for different capabilities, some stronger some weaker, but interacting as a whole and growing gradually at different paces. This is, in my opinion, the real educational and pedagogical challenge of our times—aiming to offer equal opportunities to all students according to their individual "sliders configuration".

In TE processes, teachers are aware of the differential capabilities, strengths and limitations of learners—analytic and synthetic; number oriented, word oriented or image oriented; skilled in the crafts; planners and "doers"; convergent or divergent thinkers; and soloists or collaborators. Certain design projects might offer fertile ground for the expression of all these capabilities, and this offers a further—intriguing—pedagogical challenge.

When considering diversity at a more global level, social, cultural, economic and political (SCEP) differences among communities and countries also represent significant challenges for TE. Obviously, pedagogical solutions devised for particular contexts will not necessarily be relevant in different contexts. Pedagogical solutions are means toward ends, and they reflect choices of educational philosophies (e.g., an integrated constructivist/socio-cultural stance in the proposals above; see also Chap. 8 by David Barlex, this volume). However, they are expected to serve educational goals and convey educational content in close correlation with the characteristics of specific social realities. Aiming to define the questions to be addressed in an agenda for the pedagogical future of TE, it is clear that there is a need to develop theoretical as well as practical indicators for identifying the relevant variables in a country's SCEP reality that demand particular pedagogical solutions. In addition, there is need to develop theoretical as well as practical procedures to assess the appropriateness of the suggested pedagogical solutions for the given context.

Transformations and processes at all SCEP levels in educational jurisdictions worldwide deserve the attention of researchers, decision makers and politicians, in particular in light of the astonishing pace of scientific and technological transformation. The implications of these transformations in shaping our individual and social lives, our culture and economies, unveil with increasing force the emerging differences among countries, societies and communities—that is, diversity. Education in general, but TE in particular, stands in a strong position to help address and respond to this diversity thanks to the potential embedded in technological tasks to reach all members in a community and offer them multiple ways to acquire relevant knowledge and skills.

In summary, there is no doubt that the agenda for the future should include a great deal of research and development into addressing questions related to individual, group and social diversity. In particular, I argue that it is critical that diversity is viewed as enriching, the multiple facets of a social gem, and the raw material for the creation of adaptive pedagogical solutions.

## *Emerging Pedagogies with Emerging Technologies*

These are times of rapid and continuous technological developments—the digital age. Since the sixth decade of the previous century, on-going innovations in information and communications technologies have affected education and challenged educational researchers and practitioners in search of effective as well as innovative ways to integrate digital technologies and pedagogy. The dialogue between such technology and pedagogy has, over the years, produced ideas and practices at many different levels. Salient issues include the way information is produced, stored and retrieved; the ever growing amount of information available for learning (far more than in any time in the textbook-based past), stored in varied representational formats (e.g., texts, images, pictures, video and sound clips, interactive scenarios); and the growing, rich repertoire of digital pedagogical objects (e.g., lesson plans, models and simulations, virtual environments for learning, assessment instruments, online courses). In addition, innovative pedagogical forms are being developed, studied and in cases implemented, for example, virtual gathering spaces for learning, ubiquitous-learning models, distance-learning courses (and even complete degrees), support for schooling in out-of-school settings (e.g., home, distant locations, hospitals), and developments in mobile and localised learning.

It is important to note that the new information technologies (either a physical artefact, a piece of software, or a communications-networked system) are not only a technical phenomenon. These technologies are imbued with ways of thinking about important aspects of our individual and social lives, including about the world of information, social processes and phenomena, economic processes, the workplace, and education. The implications of the impressive extent and pace of technological developments for discussion about future pedagogies are manifold. These range from supporting and upgrading current practices, to the development of new practices that were not possible before the new information technologies came to be. Examples of potential lines of research and development relevant to TE pedagogies are:

- Real-world tasks—using authentic tasks and design assignments, pedagogical formats may be developed based on in-situ data gathering and its online sharing and discussion. Mobile devices allow for data to be collected in diverse formats (e.g., pictures or video data, recorded interviews, quantitative data fed directly into appropriate software) and uploaded in real-time to public repositories for further elaboration.
- Collaborative work—public virtual spaces and shared design and analysis tools allow students to collaborate from distant locations and construct collaborative products.
- Informed design processes—information is an abundant commodity, easily accessible from any location. Herein lies significant advantages and opportunities for learning, but also some challenges. A key pedagogical question relates to the design of methods and procedures to support learners' acquisition of effective information manipulation skills. The aim is to reinforce learners' abilities to retrieve, evaluate, adapt and use information to appropriately inform the task at hand.
- Microprocessor-based tools supporting learning—many tools that are educational versions of real-world technological processing tools have reached mature status, and many more are under development (e.g., controlled manufacturing and assembly lines, robotic systems, 3D printers). The pedagogical challenge is to devise ways to integrate these (and future technologies) into constructivist pedagogies in order to transform these sophisticated tools into construction playgrounds for invention and creative learning.
- Perhaps at an even more complex and innovative level, a bigger question relates to the study of alternative formats of schooling and learning settings and configurations. For decades visions have been espoused about the possibility, enabled by the technology, to support learning beyond the constraints of space and time. Looking at the reality in most school-age systems worldwide, substantial changes have not yet taken place (Voogt and Plomp 2010). The integration of mobile devices with cloud technologies and ubiquitous wireless communication facilities are calling out to researchers and educators to explore innovative pedagogical models and conceptualisations of technology-supported teaching and learning.

Marc J. de Vries picks up many of these themes in Chap. 14, in which he argues strongly that research questions and research findings need to connect more closely with teachers. In addition, he highlights the need for research that targets how to better support students' technology learning.

## An Agenda for R&D of Pedagogical Solutions

The previous sections considered a series of issues that I believe affect, and will continue to affect, the way the pedagogy of TE is researched and implemented. Below, I suggest a conceptual framework—an R&D pedagogical ecology framework—to frame the agenda for: (a) the identification and definition of relevant research questions, and (b) the formulation of guidelines for the development of pedagogical solutions.

The R&D pedagogical ecology framework is defined on the following premises:

- There is need to base the study and development of pedagogical methods and resources in existing—and future—theoretical and research work. Many theoretical proposals have been investigated over the years, and many lessons have been learned. However, most of these insights remain confined to discussions in conference gatherings and academic publications. An agenda for the future should include an exploration of the ways in which the knowledge produced can be harnessed to inform pedagogical practices and to enrich teaching decision-making. This point is also strongly put in Chap. 14, this volume, by Marc J. de Vries.
- Pedagogical solutions should be conceived as part of a systems view of technology education in which specific (and even disciplinary) knowledge and skills are contextualised within a holistic perspective of a twenty-first century person's cognitive, affective and moral being. Pedagogical opportunities should not only lead to the achievement of specific curricular goals, but also support learners' perceptions of the gained knowledge as substantial thinking tools for interacting with real-world situations.
- Researchers and developers should be aware of the broad diversity characterising the population of learners, and the increasing need to pay attention to diversity when making pedagogical decisions. The landscape of diversity includes individual differences and capabilities; gender; social, economic, cultural or political differences; and even differences among countries' economic, cultural, technological and political realities. While the design of pedagogical resources may rely on certain commonalities, no real solution can escape the need to accommodate diversity as a critical component of different countries' realities and educational policies.

The R&D framework comprises four dimensions: Conceptual, contextual, pedagogical resources, and planning and implementation.

## *Conceptual Dimension*

The conceptual dimension refers to key conceptual components of the rationale guiding the design of pedagogical models for TE, and the planning of their systematic study. The following is but a partial account of these components, but is offered here as being indicative of the conceptual stance adopted in this chapter.

- *Foci of TE*: There is need to displace the focal point of the teaching/learning process from 'learning about' technology (still part of contemporary practice in many classrooms) to 'learning the very essence and substance' of technology. Pedagogical models should stress the importance of becoming involved in planning and implementing technological processes, coping with technology-related dilemmas, and developing awareness of the roles and implications of technology for our current and future lives.
- *Making and doing*: Consequently, pedagogies should foster both conceptual learning and praxis as necessary and complementary layers of the formation of a technologically literate person. As stated above, *to learn technology is to do technology*—in terms of conceptual and intellectual processes (e.g., analysis of and elaboration about technology-related social issues, or acquaintance with development paths of a given technology over time) *and* in terms of actual involvement in making (e.g., the design process).
- *Doing and reflecting*: Pedagogies should support and encourage students' reflection about their doing. Research observations indicate that often learners engaged in practical tasks are not requested to conceptualise and formalise the knowledge and ideas involved in, or resulting from it (e.g., McCormick et al. 1994; Rowell 2004). Thus, beyond the doing experience (valuable in itself), the distillation and abstraction of ideas and concepts is often not promoted as part of the learning process. In contrast, Stables and Kimbell (2006) have depicted students' design processes in terms of the

  > interaction between mind and hand (inside and outside the head) [...] and the activity as being best described as iterative as ideas are bounced back and forth; formulated, tested against the hard reality of the world and then reformulated. We coined the phrase 'thought in action' to summarise the idea. (pp. 315)

  Reflection, conceptualisation and explicit knowledge construction intertwined with actual making is a high priority pedagogical goal to be pursued.
- *Technological thinking toolkit*: TE should approach the teaching of concepts and skills in a way that emphasises their role as building blocks for the development of a technological worldview and technological thinking (e.g., the Nuffield 'resources' referred to in Barlex 2008, or 'primitives' in Mioduser 1998). Technological disciplines are continuously growing in scope as well as in the amount of information generated. It is critically important to equip students with a meaningful thinking-toolkit enabling them to approach, understand and consequently make decisions and act while coping with existing and future developments in the technological world. Candidate components for the toolkit might

pertain to different layers. An example at the 'rudimentary' level are schemas related to transmission configurations and parts that are found in many different artefacts, and their indexing by their advantages and constraints in different situations of use. Examples at the 'methods' layer are schemas for alternative ways to approach a technological problem and its solution.

- *Pedagogical stance*: This chapter advances the idea of a pedagogical ecology based on learning as a complex system. In this system, an intricate web of factors (related to, for example, individual capabilities, social and cultural contexts, curricular content) affect the design of specific pedagogical solutions. An essential feature in the system is the interaction between individual and cultural cognition, or more broadly between individual growth and socio-cultural processes. Thus, the pedagogical stance adopted favours constructivist, collaborative, process-oriented pedagogies over expository, "soloist", product-oriented pedagogies.
- *Values and ethical issues*: Pavlova and Middleton (2002) pointed out that "Like science, technology, and by implication, technology education, was once thought to be value-neutral. Such propositions are now discredited, however, the question concerning the values that technology educators hold is still an open one" (pp. 103). In congruity with the socio-cultural components of the rationale adopted in this chapter, pedagogical solutions should foster students' explicit confrontation with moral, ethical and value aspects of the technological processes and products under study (see also Chap. 3 by Marilyn Fleer). Authentic, situated, context-aware or prospective-evaluation activities, among others, are pedagogical resources that might contribute substantially to achieve the goal.

## Contextual Dimension

The contextual dimension of the R&D framework relates to specific contextual characteristics affecting the design of pedagogical solutions in terms of the curricular goals pursued and the populations addressed.

- *Strategic goals*: TE programmes may pursue different goals and needs, for example, technological literacy, specialisation, vocational training or professional (e.g., engineering) expertise. In general, it is seen as desirable for the whole population to acquire the rudiments and skills required to understand and act mindfully in our technology-saturated reality—to be technologically literate. In contrast, only some of the population will choose specialisation tracks in technology, usually at secondary or tertiary levels. Obviously a narrower fraction will pursue careers as expert (professional) technologists. While common pedagogical solutions can be found across all these curricular emphases, the concern here is with pedagogies aimed to support the whole population's development of technological literacy.
- *Literacy*: In this sense, the targets for the development of appropriate pedagogies are the intellectual and affective resources required to perform as technologically

literate citizens in relation to historic, current and prospective technological realities. Among these resources are: language, sets of skills and methods, a knowledge base, and the ability to evaluate and morally judge the consequences of technology-related phenomena.

- *Diversity*: In line with the above elaboration about diversity, there is need to develop and study appropriate pedagogies addressing the needs stemming from different individual, group, cultural, socio-economic or geo-political realities. At each of these levels, pedagogies should be developed targeting a range of core issues, for example, dissimilar capabilities, learning styles or interests of different individuals, and distinctive goals and needs among different population groups and even regions and countries.

## *Pedagogical Resources Dimension*

The pedagogical resources dimension of the proposed pedagogical ecology relates to characteristics of the actual resources, teaching models, and means that need to be developed and studied.

- *Knowledge and skills space*: The intense pace of technological development accentuates the need for a renewed definition of appropriate blends between different types of knowledge with regards to different age levels and targeted expertise levels. The continuously increasing body of knowledge comprises multiple layers, for example, factual, conceptual, procedural and meta-level knowledge. Pedagogical models aiming to teach the balanced blends required for acting mindfully in our changing technological world should address, besides traditional 'knowing' and 'making' skills, new sets of higher-order skills and knowledge types (DeMiranda 2004). Examples include knowledge pertaining to declarative, procedural and qualitative categories; mental modelling and running of technological systems; information manipulation skills and methods; capabilities involved in processes such as analytic, synthetic, anticipatory or goal-oriented thinking; and collaborative work skills.
- *Pedagogical repertoire*: This is the actual pedagogical toolkit comprising the building blocks out of which tasks, assignments, lesson plans, and even comprehensive curricular plans are constructed. Examples of these at the strategic level are project- or problem-based assignments, or instruction-embedded assessment tasks. Examples at the more practical level are contextual and focused methods and activities addressing specific skills or methods, in other words, candidates for modular implementation in different learning processes. As discussed above, many of these have already been developed, studied and even implemented in large curricular systems in many countries (e.g., see Barlex 2008; Compton and France 2007). A research and development agenda for the near future should consider the consolidation of previous work and further expansion and systematic study of this repertoire towards its implementation in teacher education programmes and curriculum development.

- *Classroom practices and 'the world'*: Pedagogical solutions should contribute to closing the gap between classroom practices—what is feasible within the possibilities and constraints of the classroom context—and real-world technological processes and artefacts. It is well established that many constraints (e.g., available resources, time limitations, teacher's knowledge) limit the classroom treatment of topics such as those related to complex technologies, high-tech developments, and even our sophisticated, artefact-saturated environment. Often these topics appear in 'about' boxes in textbooks or are presented in expository mode with materials from the Internet, while the actual experiences supposedly aimed to address them in class are conducted with 'low-tech' materials and processes, with students expected to make the cognitive linkages. There are cultures and societies in which there is natural flow between skills and knowledge stemming from social and economic needs, and school practices. New pedagogies should foster both classroom practices built on an essential and principled linkage with the technological processes under study, as well as situated (out-of-school and in-situ) activities in relevant technological contexts.
- *Assessment*: Existing models for assessing students' performance in technological processes (and research results of their implementation, see Kimbell et al. 2009) reinforce the need to deepen the *synergy* among the components of the teaching-learning-assessment cycle. Instead of focusing exclusively on summative accounts of learning, *process* and instruction-embedded assessment with relevance to key aspects along the learning journey should form an integral part of the pedagogies implemented. This approach stresses the importance of real-time feedback tailored to the identified needs and prospective learning paths of individuals and groups engaged in performing technological tasks (see also Chap. 7 by Kay Stables).
- *Assessment II*: Fostering the previous item's aims requires a substantial change of perspective in educational systems' perception of the interaction between teaching and assessment: from 'assessment-driven-teaching' (the prevalent policy in many educational systems) to 'learning-driven-assessment'. Pedagogies to be developed should address (and solve) the conflict between 'teaching what can be measured' versus 'measuring what can be taught' in diverse contexts and with diverse learners.
- *Learning technologies*: The rapid and on-going evolution of information technologies and the enormous effect of these on all technological fields poses serious challenges to education. Many of these new developments might function as powerful cognitive technologies when appropriately incorporated into teaching and learning processes. There is need for the design of innovative pedagogical approaches emerging from the educational use of advanced technologies, for example, ubiquitous learning, transcending the time/space boundaries of the school, collaborative and team-work supported by technology, community involvement and learning mediated by technological resources, and, at the systemic level, alternative (to school-based) learning configurations and modalities afforded by the use of technology.

## *Planning and Implementation Dimension*

As part of the elaboration about the future of the pedagogical ecology of TE, serious questions concerning policy making and systemic issues need to be addressed.

- *Flexible curricular planning*: In technology education, curricular planners and developers face the need to create the most appropriate balance between the foundational building-blocks of the discipline's knowledge base, and continuously evolving knowledge areas and fields of innovative application. Pedagogical frameworks guiding curricular development based on the printed textbook culture (characterised, for example, by structured, linear, hierarchical, within-clear-boundaries knowledge packages) have lost their relevance. Instead, new pedagogical frameworks, advancing a systems perspective about the curriculum, comprise intricate webs of interconnected knowledge (the 'hyper-curriculum') and multiple layers for approaching a topic under study (e.g., in terms of multiple sources, representation forms, or kinds of tasks—from exclusively academic to intensely hands on). Moving to these new approaches will require a wide range of theoretical and research considerations.
- *Pedagogical infrastructure*: Some of the claims presented above—such as the centrality of making and doing in TE, or the need to close the gap between technologies that are physically feasible to bring to the classroom and technologies in the world outside the classroom—require the examination of key questions concerning the pedagogical infrastructure needed to conduct tasks and projects. For example, there is no such thing as a 'technology lab' or even an appropriate resources toolkit in many educational jurisdictions, particularly at primary school level. While in some countries—but not all of them—schools have become increasingly better equipped with information technologies, these resources offer only partial answers to the pedagogical needs of TE. For example, they supply tools to access information, design software, virtual models and simulations, or control software for devices or robotic systems. More comprehensive issues related to the IT infrastructure to be incorporated in the pedagogical ecology of TE still demand thorough research.
- *Teaching configurations*: As the discipline of technology becomes increasingly complex, so does its teaching. A key aspect here is teachers' mastery of the disciplinary content as well as their pedagogical content knowledge (PCK, see Shulman 1987)—teachers' integration of specific content and pedagogical resources into sound pedagogies for particular students. More often than not, a range of expertise (rather than one teacher's focal expertise) is required to support students' learning, leading to the need for relevant team configurations. Currently this is not a reality in most educational systems, with the extreme situation occurring at the primary level where 'generalist' teachers, often not specifically trained in TE, teach this curricular area. Future research efforts should deepen the inquiry into teachers' PCK with regards to the multiple facets of the discipline, as well as team-teaching configurations and models addressing the complex pedagogical needs of TE.

- *Teacher education and professional development*: Expanding the previous point, vis-à-vis the continuous technological developments and technology-related economic and social transformations, there is urgent need to revise teachers' pre- and in-service professional development (Jones et al. 2013). In particular, it seems critical that all of the pedagogical issues raised in this chapter are addressed in all programmes designed to support educators who have to address these issues in their everyday encounters with TE students.

## Concluding Remarks

In this chapter I set out to define a future agenda for the R&D of pedagogical issues in TE from a systems perspective—TE's pedagogical ecology. Teaching and learning in TE, a complex system, comprises many interacting factors affecting both individual and social learning processes. Numerous factors, including learners, teachers, content, pedagogical means, socio-cultural realities, value systems, educational goals, labour-market expectations, organisational characteristics and institutions, and many others interact in intricate ways, affecting teaching and learning processes. I believe that this ecological/systems perspective opens a rich space of questions to be addressed in future research, as well as multiple avenues for the development and implementation of innovative TE pedagogies.

## References

Barlex, D. (1998). Design and technology: The nuffield perspective in England and wales. *International Journal of Technology and Design Education, 8*, 139–150.

Barlex, D. (2008). Nuffield Primary Design and Technology Project: A retrospective. In G. Lewis & H. Roberts (Eds.), *Design and technology in the curriculum* (pp. 37–53). Bangor: Bangor University.

Brophy, S., Klein, S., Portsmore, M., & Rogers, C. (2008). Advancing engineering education in P–12 classrooms. *Journal of Engineering Education, 97*(3), 369–387.

Cole, M., & Derry, J. (2005). We have met technology and it is us. In R. Sterneberg & D. Preiss (Eds.), *Intelligence and technology—The impact of tools on the nature and development of human abilities* (pp. 209–227). Mahwah: Erlbaum.

Collins, A. (2006). Cognitive apprenticeship. In R. Sawyer (Ed.), *The Cambridge handbook of the learning sciences* (pp. 47–60). Cambridge: Cambridge University.

Compton, V., & France, B. (2007). Redefining technological literacy in New Zealand: From concepts to curricular constructs. In J. R. Dakers, W. J. Dow, & M. J. de Vries (Eds.), *PATT 2007. Teaching and learning technological literacy in the classroom* (pp. 260–272). Glasgow: University of Glasgow.

Dakers, J. (2006). Towards a philosophy for technology education. In J. Dakers (Ed.), *Defining technological literacy: Towards an epistemological framework* (pp. 145–168). New York: Palgrave MacMillan.

Darling-Hammond, L. (2007). The flat earth and education: How America's commitment to equity will determine our future. *Educational Researcher, 36*(6), 318–334.

DeMiranda, M. (2004). The grounding of a discipline: Cognition and instruction in technology education. *International Journal of Technology and Design Education, 14*, 61–77.

Fritz, A. (1996). Reflective practice: Enhancing the outcomes of technology learning experiences. *The Journal of Design and Technology Education, 1*(3), 212–217.

Gropius, W. (1919). *Bauhaus manifesto and programme*. Weimar: The Administration of the State Bauhaus at Weimar.

Itten, J. (1975). *Design and form: The basic course at the Bauhaus and later*. New York: Wiley.

John-Steiner, V., & Mahn, H. (1996). Sociocultural approaches to learning and development: A Vygotzkian framework. *Educational Psychologist, 31*(3/4), 191–206.

Jones, A., Buntting, C., & de Vries, M. (2013). The developing field of technology education: A review to look forward. *International Journal of Technology and Design Education, 23*(2), 191–212.

Kimbell, R., Wheeler, T., Stables, K., Shepard, T., Martin, F., & Davies, D., et al. (2009). *e-Scape portfolio assessment: A research & development project for the Department of Children, Families and Schools, phase 3 report*. London: Goldsmiths, University of London.

Kolodner, J., Camp, P., Crismond, D., Fasse, B., Gray, J., Hollbrook, J., Puntambekar, S., & Ryan, M. (2003). Problem-based learning meets case-based reasoning in the middle-school science classroom: Putting learning by design into practice. *The Journal of the Learning Sciences, 12*(4), 495–547.

Lubienski, S. (2003). Celebrating diversity and denying disparities: A critical assessment. *Educational Researcher, 32*(30), 30–38.

Mawson, B. (2003). Beyond 'the design process': An alternative pedagogy for technology education. *International Journal of Technology and Design Education, 13*, 117–128.

McCormick, R., Murphy, P., & Hennessy, S. (1994). Problem-solving processes in Technology Education: A pilot study. *International Journal of Technology and Design Education, 4*, 5–34.

Mioduser, D. (1998). Framework for the study of the cognitive nature and architecture of technological problem solving. *Journal of Technology Education and Design, 8*(2), 167–184.

Mioduser, D. (2009). Learning technological problem solving—A cognitive/epistemological perspective. In A. Jones & M. de Vries (Eds.), *International handbook for research and development in technology education* (pp. 391–406). Rotterdam: Sense.

Mioduser, D., & Dagan, O. (2007). The effect of alternative approaches to design instruction (structural or functional) on students' mental models of technological design processes. *International Journal of Technology and Design Education, 17*(2), 135–148.

Nachmias, R., Mioduser, D., Dressler, M., & Mintz, R. (Eds.). (2007–2010). *New Mabat curricular project (elementary school science and technology textbooks and teacher guides series*, Hebrew and Arabic). Tel-Aviv: Ramot.

Nye, D. (2006). *Technology matters: Questions to live with*. Cambridge, MA: MIT.

O'Donnel, A. (2012). Constructivism. In R. Harris, S. Graham, T. Urdan, C. B. McCormick, G. M. Sinatra, & J. Sweller (Eds.), *APA educational psychology handbook* (Theories, constructs, and critical issues, Vol. 1, pp. 61–84). Washington, DC: American Psychological Association.

Papert, S., & Harel, I. (Eds.). (1991). *Constructionism: Research reports and essays, 1985–1990*. Norwood: Ablex Publishing Corporation.

Pavlova, M., & Middleton, H. (2002). Values in technology education: A two country study. In H. Middleton, M. Pavlova, & D. Roebuck (Eds.), *Learning in technology education: Challenges for the 21st century* (pp. 103–113). Proceedings of the 2nd Biennial international conference on technology education, Parkroyal Gold Coast, Australia.

Perkins, D. (1993). Person-plus: A distributed view of thinking and learning. In G. Salomon (Ed.), *Distributed cognitions: Psychological and educational considerations* (pp. 88–110). New York: Cambridge University.

Petrina, S. (2000). The political ecology of design and technology education: An inquiry into methods. *International Journal of Technology and Design Education, 10*, 207–237.

Prawat, R. (1996). Constructivisms, modern and postmodern. *Educational Psychologist, 31*(3/4), 215–225.

Preiss, D., & Sternberg, R. (2005). Technologies for working intelligence. In R. Sternberg & D. Preiss (Eds.), *Intelligence and technology—The impact of tools on the nature and development of human abilities* (pp. 183–208). Mahwah: Erlbaum.
Richardson, V. (2003). Constructivist pedagogy. *Teachers College Record, 105*(9), 1623–1640.
Rogoff, B., & Chavajay, P. (1995). What's become of research on the cultural basis of cognitive development? *Educational Psychologist, 50*(10), 859–877.
Rowell, P. (2004). Developing technological stance: Children's learning in technology education. *International Journal of Technology and Design Education, 14*, 45–49.
Salomon, G. (1993). No distribution without individual's cognition: A dynamic interactional view. In G. Salomon (Ed.), *Distributed cognitions: Psychological and educational considerations* (pp. 111–138). New York: Cambridge University.
Seely Brown, J., Collins, A., & Duguid, P. (1989). Situated cognition and the culture of learning. *Educational Researcher, 18*, 32–42.
Sherman, T., Sanders, M., & Kwon, H. (2010). Teaching in middle school technology education: A review of recent practices. *International Journal of Technology and Design Education, 20*(4), 367–379.
Shulman, L. (1987). Knowledge and teaching: Foundations of the new reform. *Harvard Educational Review, 57*(1), 1–22.
Sjoberg, S. (2001). *Science and technology in education – Current challenges and possible solutions*. Paper presented in the meeting of European ministers of education and research, Upsala. Retrieved at http://www.iuma.ulpgc.es/~nunez/sjobergreportsciencetech.pdf. Accessed 2 Mar 2013.
Stables, K., & Kimbell, R. (2006). Unorthodox methodologies: Approaches to understanding design and technology. In M. de Vries & I. Mottier (Eds.), *International handbook of technology education: Reviewing the past twenty years* (pp. 313–330). Rotterdam: Sense.
Voogt, J., & Plomp, T. (Guest Eds.). (2010). Innovative ICT-supported pedagogical practices: Results from the international study of information technology in education (special section: Pedagogical use of ICT worldwide, 6 articles). *Journal of Computer Assisted Learning, 26*(6), 449–522.
Walmsley, B. (2003). Partnership-centered learning: The case for pedagogical balance in technology education. *Journal of Technology Education, 14*(2), 56–69.
Williams, A., & Williams, J. (1997). Problem-based learning and appropriate methodology for technology education. *Research in Science and Technological Education, 15*(1), 91–103.

# Chapter 6
# Conversations to Support Learning in Technology Education

**Wendy Fox-Turnbull**

Quality talk plays a significant role in supporting learning in technology education. This chapter explores in depth the role talk plays in this learning in technology education and identifies three themes of talk: deployment, conduit and knowledge within it. It also explores the nature of a sociocultural perspective of thinking and learning to enhance students' achievement. Students' 'funds of knowledge' (González et al. 2005)—the knowledge and understandings gained from home and community—also play a considerable role in what students bring to, and take from, learning episodes. The chapter culminates with discussion about the implications of the above aspects of learning in technology education and explains how technology education is situated to maximise student learning in the twenty-first century.

## Sociocultural Learning Theory and Technology Education

Quality talk plays a significant role in supporting learning in technology education. This chapter explores in depth the role talk plays in this learning in technology education and identifies three themes of talk: deployment, conduit and knowledge within it. It also explores the nature of a sociocultural perspective of thinking and learning to enhance students' achievement. Students' 'funds of knowledge' (González et al. 2005)—the knowledge and understandings gained from home and community—also play a considerable role in what students bring to, and take from, learning episodes. The chapter culminates with discussion about the implications of the above aspects of learning in technology education and explains how technology education is situated to maximise student learning in the twenty-first century.

W. Fox-Turnbull (✉)
University of Canterbury, Christchurch, New Zealand
e-mail: wendy.fox-turnbull@canterbury.ac.nz

 
PJ. Williams et al. (eds.), *The Future of Technology Education*, Contemporary Issues in Technology Education, DOI 10.1007/978-981-287-170-1_6

Sociocultural theory considers the role of action and tools, including language, in the construction of knowledge (Wertsch 1998). It suggests that child cognitive development is dependent on an individual child's responses to cultural and societal influences (Resnick et al. 1991; Wertsch 1998; Wertsch et al. 1995). Sociocultural theory focuses on the role adults and/or more capable peers play in learning, with an emphasis on peer group interactions and collaborative learning (Daniels 1996; Richardson 1998). Smith (1998) suggests that from a sociocultural perspective children gradually come to know and understand the world through participation in their own activities and in communication with others. There is therefore an increasing understanding of the importance that talk plays in cognitive development (Wertsch et al. 1995).

It has long been understood that focussed and considered dialogue between teachers and their students can considerably enhance learning, although much of the dialogue between teachers and students is about management (Davis et al. 1990). There is also a considerable body of knowledge on understanding how conversations between students can enhance learning (Mercer and Littleton 2007) technology then emerges from within a social context and does not occur in isolation.

Technology is inherently socioculturally situated and value laden. Fleer and Jane (1999) argue that within a particular culture taking into consideration the social and cultural needs of the society in which it is developed. Technological knowledge includes the knowledge and understanding required to skilfully and knowledgably undertake holistic technological practice and the ability to critique existing technology and understand its complexity, including how it interacts with humans and the environment (Moreland and Cowie 2007).

Typically, in technology education classes, students are given a technological problem, communicated to them through a given brief from their teacher, for which they have to develop a technological solution. Students then engage in a selection of planned activities to allow them to develop the necessary skills and knowledge to design and possibly develop an appropriate technological solution. They subsequently undertake product development and evaluation by modelling through sketching, detailed drawing and developing three-dimensional models and/or mock-up designs. In most of this activity, oral communication is an essential ingredient whether students are working collaboratively with peers or communicating with key stakeholders. Thus, the focus on this chapter is on the nature and place of talk in the technology classroom.

## Talking for Learning

Talk is increasingly recognised as playing an important role in learning. Mercer and Dawes (2008) and Scott (2008) suggest that educational talk is either symmetrical and interactive or asymmetrical and non-interactive. *Symmetrical talk* includes verbal participation by all participants; *asymmetrical talk* involves only one person, typically the teacher or one dominant student. In traditional Western settings, much talk in the classroom is asymmetrical—teachers acting as arbiters of knowledge by leading conversations through transmission of facts, demonstrating, explaining to or correcting students. Symmetrical talk appears more congruous with recent ways of understanding learning in the twenty-first century although asymmetrical talk does still have a place in the classroom, for example, for giving instructions.

In order to understand further types of talk it is helpful to first understand argument. Mercer (2006) suggests that argument is characterised by three specific types of talk. At one end of a continuum of cognitive development is *disputational talk,* which is often asymmetrical talk, then *cumulative talk* and finally *exploratory talk*, both of which are symmetrical.

*Disputational talk* is characterised by a participant's unwillingness to understand another person's point of view with a constant reassertion of his or her own ideas. Collaborative activity becomes almost impossible as participants strive to have their views adopted. Defensive and uncooperative behaviour typify this type of talk as participants vie for power and control of the discussion.

Within symmetrical talk, as outlined in Table 6.1, cumulative talk occurs when speakers build on each other's contributions and are supportive but uncritical of other contributions. However, shared understandings are not developed and individuals retain ownership of their own understandings. In other words, there is no striving for control in cumulative talk and it does not allow for growth in shared meaning and understanding.

What Mercer (2006) identifies as *exploratory talk*, others call *dialogic thinking* (Alexander 2008) and *inter-thinking* (Mercer 2006). *Intercognitive talk* is a term introduced in this chapter to cluster the ideas represented in these types of talk. It describes conversations in which all participants contribute, learn, value, and build

**Table 6.1** Intercognitive talk

| Types of talk (How) | Explanation |
|---|---|
| Interactive and symmetrical | Verbal participation of all participants (Scott 2008). Educational talk when partners in a conversation have equal status (Mercer and Dawes 2008) |
| (i) Cumulative | Speakers build on each other's contributions, are supportive and uncritical. Shared understandings are not developed, ownership remains (Mercer 2006) |
| (ii) Intercognitive | Working collaboratively, speakers build on each other's contributions and are supportive and critical in a constructive supportive way. Participants value other contributions. Shared understandings are developed, new joint understanding develops. It includes the following: |
| | *Exploratory*: Partners engage critically and constructively in each other's ideas. Agreement is sought as a basis for joint process. Reasoning is visible in talk (Mercer 2006) |
| | *Dialogic:* Speakers are encouraged to try out new ideas. Dialogic learning demands both student engagement and intervention through talk. It draws attention from the organisational setting and concentrates on the "quality, dynamics and content of talk" (Alexander 2008, p. 10) |
| | *Inter-thinking*: Dynamic interaction of minds, joint co-ordinated intellectual activity (Mercer 2006) |

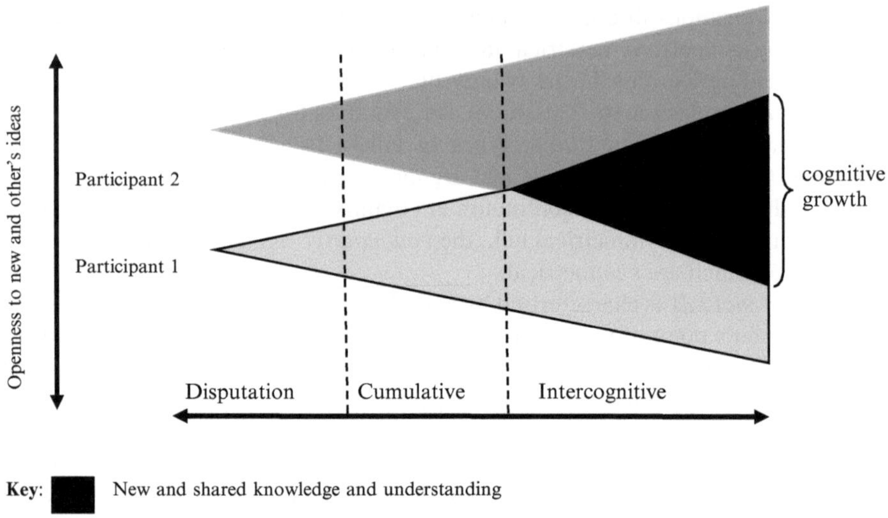


**Fig. 6.1** The relationship between shared understanding and types of talk

on each other's contributions. Participants are supportive and constructively critical of each other's contributions, and new joint understanding develops. Opinions offered, if accepted, will sway the subsequent direction of the collective thinking. Prevalence of words such as 'because', 'if', 'I think' and 'why' are used by those who use this type of talk (Mercer 2006) and can therefore be shown to be indicators of intercognitive talk.

Alexander (2008) introduces *dialogic teaching*, a pedagogical approach in which talk is given prominence. He suggests that teachers need to "provide and promote the right kind of talk" (p. 10) to ensure that students learn more effectively and efficiently. Dialogic teaching, in Alexander's view, draws attention the "quality, dynamics and content of talk" (p. 23). Mercer and Littleton (2007) identify another relevant pedagogical approach—*thinking together*—based on *inter-thinking*. Figure 6.1 attempts to show the relationship of new shared understandings generated during these three types of talk.

As understanding about the role and place of talk in learning develops, what learning or cognition actually looks like in a sociocultural setting also needs to be considered before exploring how talk can assist cogitative development in technological literacy through technological practice.

## Cognition in the Sociocultural Paradigm

Sociocultural theory, introduced by Marilyn Fleer in Chap. 3, considers the role of action and tools in the construction of knowledge (Wertsch 1998) and deals with the concept that children's cognitive development is dependent on an individual child's responses to cultural and societal influences. The goal of a sociocultural approach to learning is to understand the relationships between human action and mental functioning on the one hand, and the cultural, institutional, and historical context in which this action occurs on the other (Resnick et al. 1991; Wertsch 1998; Wertsch et al. 1995).

Murphy and Hall (2008) suggest that Vygotsky's explanation of fundamental psychological functions, such as perceptions and memory, was that they appear first as elementary functions such as a child mimicking adult behaviour by putting on adult shoes. Later, they appear as higher functions such as putting on their own shoes and understanding that shoes need to be fitting, worn on specific feet, and that different shoes are worn for different occasions. These functions occur through assimilation into sociocultural practices that are undertaken when people live and work together.

Any change in a child's development appears on two planes, first in the social plane or interpsychological functioning and then in a psychological plane or intrapsychological functioning (Murphy and Hall 2008; Rogoff and Lave 1999; Wertsch 1981; Wertsch et al. 1999). Fleer (1995) uses the example of a toddler participating in hand washing after visiting the toilet or before eating to explain these planes:

> This ritual is practised by the child's family and hence is a part of accepted behaviour patterns known to the child. However, the child may not necessarily fully understand what this action means. Vygotsky termed this social behaviour as occurring at an interpsychological level of functioning—at a social level of functioning without understanding. It is when the child understands why she/he is washing her/his hands that the child is said to be operating at an intrapsychological level of functioning. Learning occurs when the child moves from one level of functioning to another. (p. 21)

Vygotsky (1978) called the difference between a child's actual level of cognitive function and development and their potential the Zone of Proximal Development (ZPD). The role of more experienced adults or even peers is to guide children through their ZPD by modelling, talking to and challenging children into new learning. Intercognitive talk is well situated within sociocultural theory because the opposing tendencies or forces which characterise social interaction assist children to develop their own understandings and challenge others' thinking within their conversations. The next section explains how opposing ideas during interaction can assist with cognitive development in children.

## *Learning Through Interaction*

Action and mediation are two fundamental and defining themes running through sociocultural research. Mediation plays an essential role in the basic formulation of the sociocultural paradigm as it provides a link between concrete actions carried out by individuals or groups and the cultural, institutional, and historical setting in which they occur. 'Mediated action' is used very broadly to include all cooperatively and socially organised activities, inventions of shared thought (number systems, language and writing systems) and schemes for cooperative action (shared plans). It also includes a range of social rules, principles for managing recourses and relationships, and technological tools and devices (Richardson 1998). Wertsch et al. (1995) assume that action and the employed mediation exist in complex cultural, institutional and historical real world settings. These settings shape the tools when carrying out action. For example, emergence of writing has allowed the development and understanding of the structure and nature of language well beyond the original need for written communication.

The basic tenet of sociocultural conflict theory—that discrepancy or conflict best sparks cognitive development—is well positioned here. A subset of sociocultural theory, with a focus on the use of language as a tool, sociocultural conflict theory identifies conflict as an essential ingredient for any joint involvement to bring about cognitive change, thus explaining the positioning of intercognitive conversation at the centre of the 'argument' continuum. This suggests that when students' ideas are challenged and have to be defended the process can spark or enhance cognitive development. Doise and Mugny (1984) demonstrated this by showing that children working in pairs solve problems at a more advanced level than those working by themselves, regardless of the ability of the partner. They suggested that when coming up against an alternative point of view (not necessarily the correct one) in the course of joint problem solving, a student is forced to co-ordinate his or her own viewpoint with that of another. The conflict can only be genuinely resolved if cognitive restructuring takes place—in which case, mental change occurs because of social interaction. Thus the social interaction stimulates cognitive development by permitting dyadic (people working in pairs) coordination to facilitate inner coordination. This does not happen through passive presentation of points of view. When students are actively engaged in defending a particular view, and reasoning with other individuals, they experience confrontational socio-cognitive conflict. The subsequent mental restructuring allows each partner to adopt an approach to this specific class of problem that is more advanced than that adopted previously when working as an individual (Lave and Wenger 1996). Students' confidence and ability to defend their point of view can be dependent on their experiences and perceived expertise. Knowledge from their home and community, or their funds of knowledge (González et al. 2005) frequently contributes this.

## *Funds of Knowledge*

Also situated within a sociocultural paradigm, the theory of funds of knowledge draws on the perspective that learning does not just 'happen', but is a social process bound within a wider social context (González et al. 2005). *Funds of knowledge* are the developed bodies of skills and knowledge that are accumulated by a group to ensure that they can function appropriately within their social and community contexts (Lopez 2010). In Fleer's chapter in this volume, the value of funds of knowledge is clearly demonstrated by Willie, as much of what he knows was learned from the practice of his father. In the schooling context, the more teachers know about the home and cultural interests of their students the better informed they will be to maximise learning opportunities and make the most of knowledge and skills already accessible to individual students. When individual cultural knowledge is valued within the classroom, students are possibly more likely to share their knowledge with other students. Clearly talk plays an important part in this sharing, so when funds of knowledge are coupled with intercognitive conversation learning is enhanced for all involved.

The increased value of talk and the idea that students come to school with considerable knowledge and skills signals a change from many traditional education methods of the previous century. In the section below ideas for successful learning in the current century are explored.

## Learning for the Future

Gilbert (2005) suggests that new knowledge and skills are needed to enable students to succeed in the twenty-first century and to become life-long learners. Therefore, educational needs and support systems are required to re-focus the education system. Learning in the twenty-first century presents teachers with a daunting challenge of equipping students with skills and knowledge necessary to survive in the information age and beyond. Many new ideas challenge current educational assumptions and schools need to change significantly to meet the new and emerging needs of today's students. For example Claxton (2007) identify the need for greater and different student learning capacities for the twenty-first century, including students being:

- innovative, imaginative and able to problem solve
- curious, entrepreneurial and using initiative
- critical thinkers, analytical and reflective
- collaborative but also independent
- effective communicators
- resilient, determined, focused but adaptable and open-minded

It is essential that the classroom climate encourages and fosters the development of these capacities—that "students' questions are welcomed, discussed and refined, so the disposition to question becomes stronger—more and more robust; broader—more and more evident across different domains; and deeper—more and more flexible and sophisticated" (Claxton 2007, p. 120). To do this, Claxton calls for an epistemic culture change in schools to replace stand-alone courses in thinking skills or 'tricks of the trade' type learning. Aspects of this epistemic culture will include the ways teachers and learners work and talk together; the range of activities and methods they will engage in; the ways students can transfer thinking; and how teachers can model attributes, dispositions, and demeanours appropriate for successful participation in future.

Two recent curriculum developments illustrate this change in thinking. First, in New Zealand *The New Zealand Curriculum* (Ministry of Education 2007) has set a clear direction for teaching and learning in the future. There is a focus on principles such as: high expectations, inclusive bicultural (Māori and non-Māori) practices, cultural diversity, inclusion, understanding learning, community engagement, coherence and future focus, and values such as: excellence, innovation, inquiry and curiosity, diversity, equity, community and participation, ecological sustainability and integrity. The message is that, on its own, content learning in the traditional disciplines or 'subjects' will not produce the necessary skills for today's learners to be effective in the future. Second, in the United States, the *Framework for 21st Century Learning* (Partnership for 21st Century Skills 2009) sets a multifaceted direction for successful teaching and learning that includes core subjects *and* a number of life-long learning skills and dispositions including:

- Twenty-first century themes such as global awareness and financial, economic, business, entrepreneurial, civic, health and environmental literacy
- Twenty-first century learning capacities
- Technological literacy
- Life and career skills—flexibility, adaptability, initiative, self-direction, social and cross-cultural skills, productivity, accountability, leadership, and responsibility
- Twenty-first century education support systems with inclusive and varied assessment practices, instruction, professional development, and learning environments.

Inquiry Learning is one teaching approach that can successfully enable students to perform in ways described above.

## *Inquiry Learning*

Inquiry learning focuses on the facilitation of independent knowledge-based learning and reflects the belief that active involvement in construction of knowledge is essential for effective learning (Kuhlthau et al. 2007; Murdoch 2004). Inquiry learning, currently popular in New Zealand primary schools, is perfectly suited to the implementation and delivery of technology education because of the student-centred, problem-solving approach common to both. In inquiry learning, students are encouraged to construct their knowledge and understandings, enabling them to take ownership of and responsibility for learning—leading to a much broader understanding of the world. Inquiry is very different from 'open' discovery learning as teachers have a major and continuing responsibility to structure a range of activities sequenced to maximise the development of skills and thinking processes of the learners. It involves students engaging in deep learning through the process of self-motivated inquiry and strives towards developing enduring 'big understandings' and 'rich concepts' about the world and how it functions. Students develop and use, through conversation with teachers and peers, higher-order thinking skills at critical points in the learning and development process (Kuhlthau et al. 2007).

## Relevance and Implications for Technology

Aspects discussed in the previous sections of this chapter are an integral part of teaching and learning in technology education and situate it with the potential to lead education change as we move into the future. Technology education offers rich contexts for study, social construction of outcomes, connections, cooperation and collaboration with others, and practical engagement in worthwhile and real-world activities (Snape and Fox-Turnbull 2011b).

### *Technology in a Sociocultural Paradigm*

The use of culturally situated tools, including technological artefacts and language, are key factors making sociocultural theory particularly relevant to technology education. Technology is the 'know how' and creative process that may utilise tools, resources and systems to solve technological problems and enhance control over the natural and man-made environment with the aim of improving quality of life (Ministry of Education 2007). Consistent with a sociocultural paradigm, technology education sees students undertake authentic technological practice using authentic tools and practices, where learning is contextually driven.

Technology emerges from within a social context and does not occur in isolation from values, beliefs and social life. Technological outcomes are constructed within a particular culture, taking into consideration the social and cultural needs of the society in which they are developed, and those of their developers (Fleer and Jane 1999; Siraj-Blatchford 1997). I suggest that technological solutions developed within the context of the community, in which the needs arise, and those that use local skills, resources and existing technologies, are likely to be more successful than those that are not. The implication for technology education is that students must be cognisant of not only their own, but of their clients' and stakeholders' social and cultural needs when undertaking technological practice.

### *Technology as Inquiry Learning Through Immersion*

The bringing together of inquiry learning and technology education facilitates students' engagement in broad social and cultural considerations. Engaging students in meaningful contexts is essential. So too is the role of the teacher and his/her ability to be able to meet the changing and complex needs of modern teaching and learning. Authentic learning requires teachers to provide students with opportunities to understand their world and take greater responsibility using intrinsic and conative motivation (Riggs and Gholar 2009).

Technology education offers rich contexts for study, social construction of outcomes, connections, cooperation and collaboration with others, and practical engagement in worthwhile and real-world activities and authentic practices (Snape and Fox-Turnbull 2011b). Inquiry learning and technology education therefore have a number of commonalities. They are centred on both process and content, with students taking considerable ownership and responsibility (Murdoch 2004). Technology education involves the construction of technological outcomes; inquiry may similarly require the development of a tangible outcome to solve the identified problem and the development of a means of communicating the inquiry findings. Authentic inquiry practices should also be real to students, their lives, and to situations they may encounter in the future workplace. When undertaken like this, students gain an appreciation of the bigger picture (Blythe 1998; Murdoch 2004), utilise key competencies and values, create and innovate, and work with various media and educational technology. The socially embedded nature of technology integrates a variety of skills, ethics and cross-cultural themes, and technology education should provide students with opportunities to understand and participate in many local, national or global community issues. This involvement integrates a much wider range of authentic learning experiences than is offered in traditional education. Supportive and professionally aware technology teachers guide and facilitate a wide range of skills and processes; thus, their teaching can extend deep into the realm of life-long learning for successful living in the twenty-first century.

## *The Role of Conversation in Learning Technology*

Technology education projects are frequently collaborative. This requires significantly different approaches to work than the desk-confined, textbook and whiteboard techniques of traditional times. Skills required for cooperative and collaborative situations relate closely to skills and competencies identified in the *Framework for 21st Century Learning* (2009) and *The New Zealand Curriculum* (Ministry of Education 2007). The epistemic culture changes recommended by Claxton (2007) and capabilities for effective learning also link very closely to what happens in quality technology education programmes. When students work collaboratively on a single outcome or project in technology education, they must find common ground when they come across differing ideas and solutions. The quality of students' dialogue has significant impact on their ability to work and learn with and from others. Is there a rationale for the importance of dialogue beyond the fact that projects are collaborative?

To advance in technology education, students need to be taught to talk in a manner that will challenge them and their peers and allow growth and development. Although Vygotsky's work did not explicitly discuss the adult-child interactive dialogue, using the concepts of intersubjectivity and alterity can help to make sense of classroom interaction and learning that is taking place. Children's cognitive development is embedded in the context of social relationships and sociocultural tools and practices. The nature of developing technological literacy will involve varied and collaborative conversation, for example, interactions between the expert and the novice during the process of skill development, and conversations with stakeholders involving identification of needs and the presentation and justification of design ideas. When students work in teams to develop a single technological outcome, single solutions frequently have to be found to any issues. Sharing ideas and listening to each other is not enough. In order to identify a common solution or reach common ground in their thinking and move forward in the development of a successful shared outcome, students need to be able to articulate their own thinking and listen to and understand others' perspectives. This will involve intercognitive conversation—compromise and changing and developing their ideas as new material, information and thinking come to light.

A recent study investigating the nature of students' dialogue in technology education identified three distinct themes of conversation: deployment, conduit and knowledge (Fox-Turnbull 2013). The first major theme, deployment, contains conversations that show the deployment of students' existing and recently learned knowledge, either from their home, community and culture, or from knowledge learned at school—from technology education, either earlier learning or from the current topic—and from other disciplines. One significant source of 'deployment' conversations is students' funds of knowledge, rather like Willie, in Chap. 3, deploying knowledge obtained from his father to inform his knowledge and skills using metals. The second major theme, conduit, centres on the implementation of learning strategies and techniques taught, and implemented by teachers and students to manage and facilitate technology practice—thus acting as a conduit for the deployment of knowledge and experiences into contextually relevant technological knowledge and skills. The third major theme, knowledge, refers to conversations that describe the technology knowledge and skills gained during students' technological practice. These conversations are the result of a merger of deployment and conduit into student knowledge and understanding of components, practice and the nature of technology.

All themes have an interconnected relationship rather like a set of cogs. Imagine this set of three interdependent cogs, the first turning the second, which in-turn drives the third, as shown in Fig. 6.2. The increasing size of the cogs signifies cognitive growth in students as they deploy knowledge across fields of learning in a structured and planned manner such as within a purposefully-planned unit of work in technology.

The quality and effectiveness of conversation across all themes, but particularly deployment and knowledge, is enhanced by intercognitive conversation in which students listen to and contribute to each other's learning through quality dialogue.

## *Technology Education Pedagogy*

Technology is positioned to assist students' preparation for life in the current and future technological world. Barlex (2006) suggests that a major educational goal of technology is to teach students the capability to operate effectively and creatively in

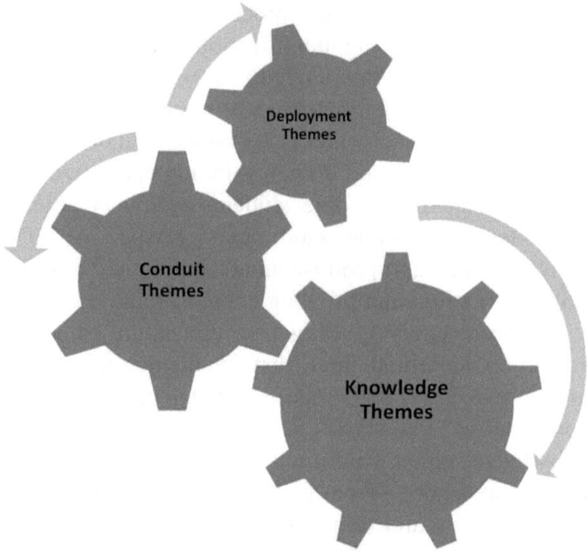


**Fig. 6.2** The interconnected nature of conversation themes (Fox-Turnbull 2013)

the made world. It must also prepare students to participate in rapidly changing technologies and to intervene creatively to improve quality of life. Teaching methods and practices must differ considerably to traditional methods used previously. For example, when designing technological outcomes, students have ownership of their design ideas and are often more knowledgeable in aspects of their practice and anyone else, including their teachers. Barlex (2006) states that from a pedagogic viewpoint this is fascinating—it is the pupil who has the knowledge and expertise in this situation; only he/she knows about his/her design.

This shift in learning outcomes requires quite a different role and approach by teachers than previously. Teachers and other experts must facilitate students' learning to enable progression of the intended design and develop knowledge and understanding of the wider social, cultural, ethical and environmental considerations that impact or influence the design. It will include critical reflection and feedback through a range of strategies and activities to motivate, engage, develop and challenge students. This approach involves the giving and receiving of constructive criticism in a supportive environment that is conducive to change and growth.

## *Dimensions of Authenticity in Technology Education*

Authenticity in technology education occurs through specific links to students' context and real technological practice. This definition is predominantly based on connecting students' understanding to meaningful and real-world situations and their involvement in technological practice that is similar to practicing technologists, using authentic tools and processes where possible. Hennessy and Murphy (1999) explain that authentic practice involves situations that are real to the student, their lives, and to situations they may encounter in the future workplace. Activity embedded in authentic technological practice is more likely to produce greater understanding and provide the opportunities for students to identify simulate and relate to the tacit knowledge of technologists. Snape and Fox-Turnbull (2011a) suggest that three dimensions of authenticity enhance technology education: pedagogy and instruction, teachers and learners, and activities. These three dimensions are brought together in complex and rich tasks, interaction with technological communities of practice, using cognitive and metacognitive instruction and thinking, the affective and emotional aspects of learning, active and collaborative participation, and dealing with meaningful problems and issues.

Newmann and Wehlage (1993) base authentic achievement on three criteria: students will construct meaning and produce knowledge, this knowledge is achieved through disciplined inquiry, and students will work toward production of discourse, products, and performances that have value or meaning beyond success in school. Slavkin (2004) identifies that learners function best in environments that are intriguing, multi-sensory and dynamic. Real-world, rich problems provide the opportunity for collaboration and the high-level discourse required for deeper learning. Interactivity between the student and the wider community is fundamental to shifting the focus of learning away from the teacher. Learning should also closely resemble everyday situations, providing students with opportunities to make decisions about the nature, content and pace of their learning (Petraglia 1998).

'Authentic' teachers take responsibility for keeping up-to-date and aware of the variety of possible opportunities that exist for student involvement and engagement (Kreber et al. 2007). Teachers need to integrate aspects of key competencies and values into their subject areas and, as professionals, teachers must ensure that their teaching pays particular attention to what is best for the students and their understanding, to help them make better sense of the world in which they live. Cranton (2001, cited Kreber et al. 2007, p. 34) argues that "the authentic teacher cares about teaching, believes in its value, wants to work well with students, and has a professional respect for students". Kreber et al. state that authentic teachers:

- engage with larger questions of purpose;
- convey how their subject matter matters in the real-world;
- connect learners in substantive authentic conversations or dialogue around significant issues; and
- are guided more by caring for the education of students than by their own self-interest.

Student's ownership and motivation to engage (Murdoch and Hornsby 2003) is necessary if enduring learning is to take place. Riggs and Gholar (2009) focus on the role that students can themselves play to accomplish their dreams and aspirations. They describe this as the conative domain, or conation—the will, drive or determination to achieve a goal. Riggs and Gholar state that "the conative connection focuses on two objectives. Firstly *knowing* what one has to do to achieve a special goal and secondly *doing* what one has to do, intentionally giving one's personal best to achieve a specific goal" (p. x, original emphasis). They identify the fundamental attributes of conation as belief, courage, energy, commitment, conviction and change. Others have referred to similar aspects as self-actualisation, self-efficacy or individuation (Kreber et al. 2007; Tessmer and Richey 1996).

A strong link to intrinsic motivation develops through the challenge, relevance, interest and involvement in the contexts students study:

> The learner must choose to learn and the learner must have enough courage to make the choice to learn. Making the choice to learn is influenced by the learner's perception of herself, her perception of the world around her, her beliefs, how she interprets what she knows or thinks she knows and how she chooses to respond to what she believes (Riggs and Gholar 2009, pp. 46–47)

Empowered students given choice, responsibility and encouragement will mostly flourish and develop the skills and frameworks needed for successful authentic and life-long learning. For these students, motivation, drive and determination will not be problematic as they learn effectively and can successfully relate to connections in the world in which they live.

Reeves et al. (2002) present ten design characteristics of authentic activities identified in literature and suggested they could make a suitable checklist for educators. Such authentic activities:

- have real-world relevance;
- are ill-defined requiring students to define them in order to complete the activity;
- comprise complex tasks to be investigated by students over a sustained period of time;
- provide the opportunity for students to examine the task from different perspectives, using a variety of resources;
- provide the opportunity to collaborate;
- provide the opportunity to reflect;
- can be integrated and applied across different subject areas and lead beyond domain-specific outcomes;
- are seamlessly integrated with assessment;
- create polished products valuable in their own right rather than as preparation for something else; and
- allow competing solutions and diversity of outcome.

## *Conation in Technology Education*

Technology education, through the development of student-driven technological outcomes, is well situated to develop conation within students. By being cognisant of the dimensions of authenticity, and when woven together by four critical aspects of twenty-first century learning, authentic technological practice can be enhanced and teachers should be able to foster high levels of conation in students. The four critical aspects of learning include: rich contexts, social construction, connections

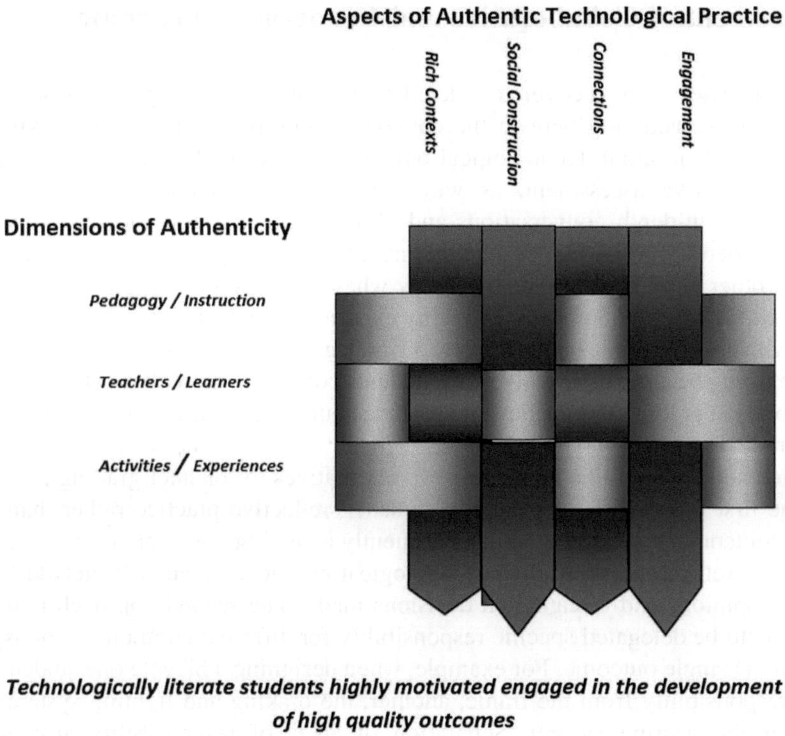


**Fig. 6.3** A model of authentic technology for producing quality technological outcomes (Adapted from Snape and Fox-Turnbull 2011a)

and engagement. The development of rich contexts with real, substantive and complex problems enabling significant developmental, intellectual and cognitive growth (Blythe 1998) enables the *social construction* of meaningful *connections* that facilitates an outcome student of *engagement*. The end result is technologically literate students who are highly motivated and engaged in the development of high quality outcomes. This notion is illustrated in Fig. 6.3.

## *Collaborative Learning in Technology Education*

To model collaborative and cooperative technology practice and prepare students for a future in technology, students should be required to work with peers and seek the opinions of experts and stakeholders from the wider community. In addition, they need to be able to alter their own views as new knowledge comes to light. For a group of students to be able to work collaboratively and cooperatively on the development of single technological outcomes, clear communication and ultimately consensus is essential. Students need to be able to discuss, debate, disagree and reason with open minds to solve the technological problems they encounter.

## Assessment of Collaborative and Cooperative Learning

Although assessment is covered in detail by Kay Stables in Chap. 7 of this volume, it is worth considering here in the context of multiple students undertaking the development of single technological outcomes. Issues of fairness arise primarily with summative assessment, as with formative assessment identified success criteria and in-depth conversations and observations can assist teachers and students to identify individual current and next step learning. Summative assessment on the other hand is different, especially when students are being graded and / or assessed for external purposes. From my experience students often feel uncomfortable with a single grade being given to a whole group, regardless of individual input and learning because invariably some students contribute more than others do. I call this 'blanket grading' and believe it can and should be avoided. Even when working on single outcomes, no two students will have exactly the same technological practice and I would like to suggest two alternatives to 'blanket grading'.

The first is to assess individually students' reflective practice rather than their actual outcome, by ensuring and subsequently assessing a written, oral or pictorial record of reflections about their technological practice, which will include justification, opinions and thoughts on decisions made. The second approach is for the students to be delegated specific responsibility for different components or aspects within the single outcome. For example, when designing a bicycle one student may take responsibility from the frame, another, the braking and lighting systems and another the gearing system. Separation of areas of responsibility also allows individual assessment of specific aspects of an outcome.

## Moving Forward in Teaching Technology—Where to Next?

To learn technology effectively students need to be engaged in contexts that are future focussed, real, substantive, and complex. The contexts must enable significant intellectual and cognitive growth and development. They will relate to the immediate and future worlds of the students and facilitate development of key competencies or dispositions identified as essential for learners. Students must not only be involved in the development of technological outcomes but should also be involved in critique of the technological world. Contexts should offer multiple solutions and include breadth in experiential learning. The learning contexts will promote big understandings about the world in which students learn, connecting threads across a range of contexts—such as sustainability, competition or cyclic resources—and how learning is situated in relation to these (Blythe 1998).

Fleer and Quiňones (2009) state that to understand and to be able to assess children's technological knowledge teachers need to understand the social and cultural context of their learners. The knowledge that students come to school with can enhance their learning and facilitate useful interactions between knowledge found inside and outside the classroom. They suggest that teachers can make more of the learning in their classrooms if they understand that students bring with them knowledge from their families, culture and background and that teachers can legitimise this knowledge through purposeful classroom engagement, "one can create conditions for fruitful interactions between knowledge found inside and outside the classroom" (González et al. 2005, p. 20). This is particularly relevant to technology education as students will often bring knowledge of technological artefacts, systems and processes to their school-based technology practice. Learning starts from a student's existing knowledge. Having identified this prior knowledge, teachers are able to engage and motivate students by selecting relevant and culturally appropriate technological contexts for learning. When teachers have the mind-set that all students can improve and that students bring their culture and experiences to learning they are in a good position to plan effective, engaging learning (Clarke 2008; Harrison 2009).

## *Clarity of Learning*

Teachers need to become conversant with formative assessment (Black and Wiliam 1998) and the principles of active learning (Newmann and Wehlage 1993), inquiry learning (Darling-Hammond 2008) and collaborative learning (Brown and Thompson 2000) to increase students' achievement. Clarke (2005) supports the need for students to take a greater role in their learning, through the use of learning intentions, questioning, self and peer assessment, and the use of formative assessment. This involvement increases as students become more aware of the purpose, intent and scope of their learning. Teachers explicitly identifying learning intentions will tend to focus more on actual learning than the activity or experience being used to stimulate student thinking. When appropriate success criteria for the learning are highlighted or co-constructed, the learning will be even greater.

## *Identification and Articulation of Explicit Learning Outcomes*

The identification of clear learning intentions in technology education will inform students which aspect of technological knowledge and skills is to be the focus of learning within any one lesson. Teachers need to consider the knowledge (conceptual, procedural, and societal) and skills (technical and information) students will need to enable relevant learning. Each learning intention should be associated with a planned experience to facilitate that learning. Learning experiences must be purposeful and logically sequenced. This ensures students have enough relevant information about both the context of their study and the necessary technological knowledge and skills to enable the development of their intended outcomes. They should also assist students to understand the societal, environmental and global issues that may affect their decision-making and final designs.

Each learning intention and associated experience usually produces some form of tangible (written, oral, visual, dramatic, graphical) evidence of learning and is able to be used formatively by students and teachers for assessment. They may include such things as posters, charts, interviews, written summaries or reviews, products systems or environment plans, discussion or oral explanations, concept maps, annotated drawings, 2D and 3D planning, functional and prototype models and or the outcome. Evidence that the predetermined learning outcomes have been met can be used formatively or summatively when clear criteria have been identified.

## Identification of Context Free Learning

When writing learning intentions for specific technological knowledge and skills it is easy to muddle context and technological learning, so that neither become clear—developing 'mucky brown paint' rather than a clear pool of 'curriculum colour' is a risk, with technological learning buried in 'busy activity' related to the context (Fox-Turnbull 2012). The separation of learning objective and context ensures that students and teacher clearly focus on technological learning. Clarke (2008) suggests separating context from learning intention can have a dramatic effect on teaching and learning. Context—the activity or "vehicle" through which learning occurs—is, however, vitally important. Examples of learning intentions that have been separated from the learning context are given below in Table 6.2. The context-free learning intention, when articulated to students, enables both teachers and students to focus on the key technological concepts and ideas to be taught and subsequently assessed.

**Table 6.2** Learning objective with context separated (Compton 2011)

| Learning intention | |
|---|---|
| Students are learning to..... | Context |
| Understand the importance making a mock-up has on the quality of a final outcome | School senior ball gown/prom dress |
| Plan technological practice through the development of a critical path to maximise all team members' use of time | Meals for the elderly living at home |
| Understand how the physical and functional nature of an outcome impacts on performance | Wooden jigsaw puzzles for an early childhood centre |
| Explore critical environmental issues and impacts of a commonly used technology | Cell phones |
| Draw 3D detailed plans of a structure using a suitable software programme | Hutch for a rabbit |
| Construct quality, safe, user friendly technological outcome for a specific client | Webpage for a local institution/club |

Making learning objectives and the separated contexts very clear to students may support them to transfer skills and knowledge from other contexts within and across the curriculum (Clarke 2008). However some learning intentions within any unit of work will be focussed specifically on the context, for example, a historical study of ball gowns or understanding the stages of child development enhanced through doing jigsaw puzzles. Again, the learning intention must be clearly articulated to students so that they may understand the purpose of their learning.

## Conclusion

As we move further into the twenty-first century, with the rapid pace of technological change, technology education offers a unique opportunity to develop understanding of the power and place of talk in learning, and the nuances of inquiry learning. This chapter introduces three main sources of talk that contribute to students learning in technology. It also argues that there are numerous similarities between inquiry learning and technology education. I suggest, therefore, that they are perfect partners. Sociocultural theory, inquiry learning and twenty-first century learning all situate the learner at the centre of learning. Quality technology education programmes do the same. Learning well informed by teacher knowledge and theory of authentic learning also offers insight into the potential of collaborative, practically-based curriculum areas such as technology education and, because interaction enhances learning in student-based approaches, the place and role of talk in learning technology is critical to support students' understanding in technology education.

Implications for technology teachers and students are numerous. Teachers of technology are in a critical position to assist students for future success. However, in order to do this, they must let go of control and the need to be the "leading light" or "fountain of all knowledge" in the classroom and empower their students to take control of their own learning. Teachers also need to determine which knowledge and skills need to be taught to whole class groups as 'basics' and which are taught on a need to know basis. When and how to influence and assists students' practice is another important consideration for teachers. Too much intervention too soon may be disempowering for students; insufficient assistance or intervention too late may see students giving up or disengaging.

Students also may need to change. Quality technology education programmes based on widely agreed principles of learning will enable students to drive their own learning. To do this they need to demonstrate conation, be motivated and engaged in their learning, talk to their peers and teachers, be prepared to have their ideas debated, and be open to others' ideas. Learning outside the classroom will play an important role and students need to learn to make connections between multiple aspects of their life and experiences.

Technology education is not only well situated in the curriculum to support students' inquiry learning, but also has the potential to demonstrate the success of an inquiry approach to learning for all teachers and students as they come to grips with recent movements in the understanding of teaching and learning.

## References

Alexander, R. (2008). *Towards dialogic teaching: Rethinking classroom talk* (4th ed.). Cambridge: Dialogos.

Barlex, D. (2006). Pedagogy to promote reflection and understanding in school technology courses. In J. Dakers (Ed.), *Defining technological literacy—Towards an epistemological framework* (pp. 179–196). New York: Palgrave MacMillan.

Black, P., & Wiliam, D. (1998). *Inside the black box—Raising standards through classroom assessment* (1st ed.). London: King's College.

Blythe, T. (1998). *The teaching for understanding guide*. San Francisco: Jossey-Bass.

Brown, D., & Thompson, T. (2000). *Cooperative learning in New Zealand schools*. Palmerston North: Dunmore.

Clarke, S. (2005). *Formative assessment in action: Weaving the elements together*. Oxon: Hodder Murray.

Clarke, S. (2008). *Active learning through formative assessment*. London: Hodder Education.

Claxton, G. (2007). Expanding young people's capacity to learn. *British Journal of Educational Studies, 55*(2), 115–134.

Compton, V. (2011). Technology in the primary sector in New Zealand: Reviewing the past twenty years. In C. Benson & J. Lund (Eds.), *International handbook of primary technology education* (Vol. 7, pp. 29–38). Rotterdam: Sense.

Daniels, H. (1996). *An introduction to Vygotsky*. London: Routledge.

Darling-Hammond, L. (2008). *Powerful learning*. San Francisco: Jossey-Bass.

Davis, R., Mahar, C., & Noddings, N. (Eds.). (1990). *Constructivist views on the teaching and learning of mathematics*. Reston: National Council for Teachers of Mathematics.

Doise, W., & Mugny, G. (1984). *The social development of intellect*. Oxford: Pergamon.

Fleer, M. (1995). *Staff-child interactions—A Vygotskian perspective*. Canberra: Australian Early Childhood Association Inc.

Fleer, M., & Jane, B. (1999). *Technology for children: Developing your own approach*. Erskineville: Prentice Hall.

Fleer, M., & Quiñones, G. (2009). Assessment of children's technological funds of knowledge as embedded community practices. In A. Jones & M. J. de Vries (Eds.), *International handbook of research and development in technology education* (pp. 477–491). Rotterdam: Sense.

Fox-Turnbull, W. (2012). Learning in technology. In P. J. Williams (Ed.), *Technology education for teachers* (pp. 55–92). Rotterdam: Sense.

Fox-Turnbull, W. (2013). *Themes of conversation in technology education*. Paper presented at the International Technology and Engineering Educators' Association Conference, Columbus, OH.

Gilbert, J. (2005). *Catching the knowledge wave? The knowledge society and the future of education*. Wellington: NZCER.

González, N., Moll, L. C., & Amanti, C. (Eds.). (2005). *Funds of knowledge*. New York: Routledge.

Harrison, C. (2009). Assessment for learning. In A. Jones & M. de Vries (Eds.), *International handbook of research and development in technology education* (pp. 449–459). Rotterdam: Sense.

Hennessy, S., & Murphy, P. (1999). The potential for collaborative problem solving in design and technology. *International Journal of Technology and Design Education, 9*(1), 1–36.

Kreber, C., Klampfleitner, M., McCune, V., Bayne, S., & Knottenbelt, M. (2007). What do you mean by "authentic"? A comparative review of the literature on conceptions of authenticity in teaching. *Adult Education Quarterly, American Association for Adult and Continuing Education, 58*(1), 22–43.
Kuhlthau, C., Maniotes, K., & Caspari, A. (2007). *Guided inquiry: Learning in the 21st century.* Westport: Libraries Unlimited.
Lave, J., & Wenger, E. (1996). *Situated learning: Legitimate peripheral participation.* Cambridge: Cambridge University.
Lopez, J. K. (2010). *Funds of knowledge. Learn NC.* Retrieved from www.learnnc.org
Mercer, N. (2006). *Words & minds: How we use language to think together.* Abingdon: Routledge.
Mercer, N., & Dawes, L. (2008). The value of exploratory talk. In N. Mercer & S. Hodgkinson (Eds.), *Exploring talk in school* (pp. 55–71). London: Sage.
Mercer, N., & Littleton, K. (2007). *Dialogue and the development of children's thinking—A sociocultural approach.* Oxon: Routledge.
Ministry of Education. (2007). *The New Zealand curriculum.* Wellington: Learning Media.
Moreland, J., & Cowie, B. (2007). Teaching approaches. In M. de Vries, R. Custer, J. Dakers, & G. Martin (Eds.), *Analyzing best practices in technology education* (pp. 213–219). Rotterdam: Sense.
Murdoch, K. (2004). *Classroom connections—Strategies for integrated learning.* South Yarra: Eleanor Curtain.
Murdoch, K., & Hornsby, D. (2003). *Planning curriculum connections. Whole school planning for integrated curriculum.* South Yarra: Eleanor Curtain.
Murphy, P., & Hall, K. (2008). *Learning and practice: Agency and identities.* London: Sage.
Newmann, F. M., & Wehlage, G. G. (1993). Educational leadership: Five standards of authentic instruction. *Educational Leadership, 50*(7), 8–12.
Partnership for 21st Century Skills. (2009). *A framework for 21st century skills.* Accessed 1 Feb 2014 from www.p21.org
Petraglia, J. (1998). The [mis]application of constructivism to the design of educational technology. The real world on a short leash. *Journal of Educational Research and Development, 46*(3), 53–65.
Reeves, T., Herrington, J., & Oliver, R. (2002). Authentic activities and online learning. In A. Goody, J. Herrington, & M. Northcote (Eds.), *Quality conversations: Research and development in higher education* (Vol. 25, pp. 562–567). Jamison: HERDSA.
Resnick, L. B., Levine, J. M., & Teasley, S. D. (Eds.). (1991). *Perspectives on social shared cognition.* Washington: American Psychological Association.
Richardson, K. (1998). *Models of cognitive development.* Hove: Psychology Press Ltd.
Riggs, E. G., & Gholar, C. R. (2009). *Strategies that promote students engagement* (2nd ed.). California: Corwin.
Rogoff, B., & Lave, J. (1999). *Everyday cognition: Its development in social context.* Cambridge, MA: Harvard University.
Scott, P. (2008). Talking a way to understanding in science classrooms. In N. Mercer & S. Hodgkinson (Eds.), *Exploring talk in school: Inspired by the work of Douglas Barnes* (pp. 17–36). London: Sage.
Siraj-Blatchford, J. (1997). *Learning technology, science and social justice: An integrated approach for 3–13 year olds.* Nottingham: Education Now.
Slavkin, M. L. (2004). *Authentic learning: How learning about the brain can shape the development of students.* Maryland: Scarecrow Education.
Smith, A. B. (1998). *Understanding children's development* (4th ed.). Wellington: Bridget Williams Books Ltd.
Snape, P., & Fox-Turnbull, W. (2011a). Perspectives of authenticity: Implementation in technology education. *International Journal of Technology and Design Education, 23*(1), 51–68.
Snape, P., & Fox-Turnbull, W. (2011b). Twenty-first century learning and technology education nexus. *Problems of Education in the 21st Century, 34*, 149–161.
Tessmer, M., & Richey, R. C. (1996). The role of the context of learning and instructional design. *Journal of Educational Technology Research and Development, 45*(2), 85–115.

Vygotsky, L. S. (1978). *Mind in society: The development of higher psychological processes*. Cambridge, MA: Harvard University.
Wertsch, J. (Ed.). (1981). *General genetic law of cultural development*. Armonk: Sharp.
Wertsch, J. (1998). *Mind as action: The task of sociocultural analysis*. New York: Oxford University.
Wertsch, J., Del Rio, P., & Alverez, A. (Eds.). (1995). *Sociocultural studies of the mind*. Cambridge: Cambridge University.
Wertsch, J., Minick, N., & Arns, F. (Eds.). (1999). *The creation of context in joint problem-solving*. Cambridge, MA: Harvard University.

# Chapter 7
# Assessment: Feedback from Our Pasts, Feedforward for Our Futures

**Kay Stables**

This chapter begins by pointing out that the behaviourist paradigm that still largely dominates assessment practices conflicts with more progressive understandings of constructivist and sociocultural approaches to learning and teaching. In order to progress assessment in Technology Education, Kay argues for adopting a pedagogic approach to assessment (where both teaching and assessment practices aim to support learning), maintaining authenticity in activities through which assessment is being undertaken, recognising the importance of judgement in valid processes of assessment, and maintaining a focus on equity and the inclusive role of the learner. The chapter also considers the potential affordances of new technologies in assessment. The chapter concludes by pointing out that future developments should support teachers to align learning and assessment. This is to ensure that learners engage in technological practice that makes visible to them and their teachers and assessors the learning that has taken place and the capability that has been developed.

## Introduction

Some years ago, on behalf of *Design and Technology: An International Journal*, I was guest editor for an issue of the journal on the topic of assessment. In introducing the theme of the special edition, I drew an example of learning from my personal life that highlighted some important issues about assessment, and I'm taking the liberty of repeating the example here as a starting point from which to explore the current state of play in assessment and considerations for how future agendas might be developed. The example reflects on my experience of buying a new washing machine.

K. Stables (✉)
Design Department, Goldsmiths, University of London, New Cross, London, SE14 6NW
e-mail: k.stables@gold.ac.uk



PJ. Williams et al. (eds.), *The Future of Technology Education*, Contemporary Issues in Technology Education, DOI 10.1007/978-981-287-170-1_7

> One of the realities of learning processes is that you only know that, or how well, you have learned something if there is some kind of feedback in the system. I know that I have learnt how to make my new washing machine work when I put in dirty clothes, and some time later take out clean ones. I also get a whole series of clues about my success along the way: I turn a dial and a light indicates that I have selected a particular wash programme; I re-set the dial and find that I can choose between more than one programme; a further light indicates which button has to be pressed to start the process—and when the button doesn't work another one indicates that I have yet to close the door. Then comes the reassuring sound of water flooding into the machine . . . I have succeeded in getting the process off the ground. The feedback I am receiving is doing two things. It is reassuring me when I get things right and it is providing clues or prompts to help me learn. If I get really stuck, there is always the instruction book to refer to, which gives more detailed and authoritative advice, although often far more than I need at that particular moment.
>
> So, with good feedback, a sprinkling of problem solving, a level of personal confidence and an authoritative source to turn to when needed, my learning in respect of using a washing machine makes good progress. What is notable though, is that before I engaged in this process, no one assessed me to see if I was capable of using a washing machine and at the end no one assessed me to see what standard in using washing machines I had reached. And yet the learning took place. (Stables 2007, p. 3)

What might be seen as the implication of this example is that we don't need assessment to learn. If this is the case, why dedicate a whole book chapter to the topic? But the example is set in a very particular context—the learner had real motivation to learn, a genuine 'need to know' in order to get the washing clean. The learner was also a confident adult with considerable life experience (including of washing machines) to draw on, already had a repertoire of approaches to problem solving and self discovery, and was receiving feedback 'prompts' from the machine itself. In many learning situations things are not quite so benign. And in many assessment situations, supporting learning is not a priority. The name of the game is labeling and sorting: can use a washing machine; can't use a washing machine. For some, 'good' assessment is that which is supporting the development of a learner. For others, 'good' assessment is that which allows us to reliably categorise people by what they can or can't do, or know or don't know. These different standpoints can be witnessed in education systems and settings across the globe and can be extremely problematic when trying to agree on valuable, effective systems of assessment. So how did we get to where we are?

It has been recorded that formal written tests go back over 2,000 years. Introduced in China, written tests were created to provide a meritocratic route into becoming a civil servant—"for identifying the talented among the common people" (Hanson 1994, p. 186). The history of these very early assessments is fascinating but for our purpose it is interesting to note, briefly, the age-old assessment issues that Hanson's history displays. Parental involvement in preparing young people for the test even went so far as pregnant mothers being exposed to test-relevant content. Young people 'crammed' for the tests by being locked in a room for a year. So many took the tests that they were locked in cells for 3 days to complete them, first having been rigorously searched to avoid cheating and checked to see that the person wasn't an imposter. Extreme lengths were employed to ensure candidates' submissions were anonymous. The tests themselves were so focused on standardised results based on correct responses that, despite their meritocratic underpinning of seeking talent in 'common people', this standardisation was what finally brought the system down. It stifled creativity and resulted in an assessment system that didn't provide the evidence that was required—to know whether the person tested was qualified to become an effective civil servant.

While being quite extreme, it is surprising how the spirit of this early approach still haunts us today. Kelly (2009) cites the Taunton Report (1868) and the Beloe Report (SSEC 1960) in reminding us that "[t]he assessment (and evaluation) tail will always wag the curriculum dog" (p. 148). We have also seen in Technology Education how assessment has distorted the curriculum (e.g., Atkinson 2000; Harris and Wilson 2003; Kimbell 1997, 2006; OfSted 2001, 2002). Even the overarching aim of a curriculum can be undermined, as was seen in the Prest (2002) review where, for example, an aim of the English National Curriculum for Design and Technology (D&T) was for learners to "think and intervene creatively" but the assessment criteria made no mention of creativity, resulting in it being undervalued and under-prioritised in classroom activities.

## Different Purposes, Different Approaches

### *Assessment of Learning or Assessment for Learning?*

Much of the history of assessment is based in what Hanson (1994) refers to as "qualifying tests" that exist to see if a person is fit to move into a particular place within society and also within a specific education system. In education, the history has been of the development of qualifying tests as 'entrance exams', for example, to gain a place in University. These entrance exams soon spawned the development of 'exit' exams, such as school leaving exams, which created a chain of 'qualifying tests': take an entrance exam to see if you qualify for a school place, take an exit exam at school to see if you qualify for a university place, and so on. Such assessment systems are ubiquitous in school systems across the globe and across disciplines, including Technology Education. Described by MacLeod (1982) as "the single most intrusive and expensive innovation in Western education" (p. 16), they are often accompanied by disturbing effects, as Hanson describes:

> regardless of people's feelings about them, qualifying tests are a key factor for living successfully in contemporary society. Those who reject the message of personal insufficiency reiterated by poor test performance may turn off on tests, but then the system turns off on them. They are excluded from educational opportunities and good jobs and (just as the tests predicted!) they never are able to accomplish much. . . . Qualifying tests constitute one of the central conditions of contemporary society. (p. 186)

The contrast between my *personalised* learning in the washing machine example and Hanson's analysis of effects of a history of qualifying tests could not be more stark and yet they can both be viewed on an assessment continuum as, for example, characterised by Harlen and Deakin Crick (2002): "Assessment is a term that covers any activity in which evidence of learning is collected in a planned and systematic way, and is used to make a judgment about learning" (p. 1).

The term assessment has now become so common and embedded in our education systems that there is a danger that a single view is assumed of what it is taken to mean. The reality is that the term has many meanings, behind which lie a plethora of different functions of assessment and different philosophies of learning.

Much has been written about the purposes of assessment. A useful and straightforward lens through which to view these is provided through the extension of Harlen and Deakin Crick's (2002) definition used above, in which they dichotomise the function of assessment:

> If the purpose is to help in decisions about how to advance learning and the judgement is about the next steps in learning and how to take them, then the assessment is formative in function. If the purpose is to summarise the learning that had taken place in order to grade, certificate or record progress, then the assessment is summative in function. (p. 1)

Assessment purposes can therefore be seen on a continuum. At one end sit the formal 'qualifying tests' described by Hanson (1994)—assessment *of* learning. At the other end sits my washing machine example—assessment *for* learning, assessment to support the learning as it is taking place.

## *Written or Practical?*

When we turn to a curriculum area such as Technology Education, the territory of assessment becomes further complicated by the practice-based nature of the subject. The examples of qualifying tests alluded to earlier are largely of written tests. Yet early examples of qualifying tests of practice also exist, for example, those initiated in mediaeval times that were the basis of qualifying within the guild of a particular craft through an apprenticeship model of learning. Assessment in Technology Education has developed from both roots—the qualifying written test and the qualifying practical test. But the roots of each grow in different soil, the written tests coming from an academic tradition where 'knowledge' can be demonstrated through the written word, the practice-based tests coming from what Fleming (2013) describes as a holistic model where learning is through mentoring and direct experience and where 'knowledge' is embedded in the quality of material artefacts. These dual systems have presented conflict and challenges for Technology Education, exacerbated by the seemingly higher status of the written test and its link to an academic education and the comparatively lower status of the practical test and its link to vocational education. To bring further understanding to the complexities in assessment, we turn next to who assessment is for—who are the stakeholders in the process?

## *Stakeholders in Assessment*

The range of stakeholders in educational assessment systems is diverse, including learners, teachers, parents and guardians, school administrators, policy makers, politicians and employers. With each stakeholder group come different sets of understandings, values and agendas. The world of assessment can be presented as being as simple and straightforward as Harlen and Deakin Crick's (2002) clear categorisation might appear to suggest, but the reality is likely to be far more complex and messy.

Taking into account the stakeholders in any assessment situation adds complexity, increased still further when we consider how different stakeholders perceive feedback from assessments, and the potential consequent actions (the feedforwards) that are conceived. If the teacher is the stakeholder in formative assessment, then the purpose may become diagnostic, as they gain better insight into the learner's understanding, ability, skill, etc. and plan the next step in learning. If the learner is the stakeholder, the purpose may be metacognitive, as the learner comes to have a better understanding of her or his own understanding and how this can be used in future situations. Summative assessment for a teacher may be evaluative as he or she reflects on the learning experiences that have been provided and how these experiences might be modified in the future. For a policy maker, however, the feedback from summative assessments may be read quite differently, for example, resulting in decisions on how the teaching workforce is paid.

Historically there has been a dominating thrust, particularly in Western education, for assessment to be focused on the summative function of producing grades and certification as evidence of 'exit' qualifications to inform the gatekeepers of future education or employment. By prioritising in this way, assessments create extrinsic motivation for the learner under scrutiny. However, recent decades have seen a swing of the pendulum to more pedagogic approaches to assessment where the primary focus has been to support the processes of learning. An example of this swing can be seen in policy documentation that supported the initial development of a National Curriculum (NC) in England and Wales in 1990. From the report of the Task Group on Assessment and Testing (the TGAT report, DES/WO 1988) led by Professor Paul Black, we have a clear statement of this shift:

> Promoting children's learning is a principal aim of schools. Assessment lies at the heart of this process. It can provide a framework in which educational objectives may be set, and pupils' progress charted and expressed. It can yield a basis for planning the next educational steps in response to children's needs. (para 3)

This statement illustrates a paradigm shift to a pedagogic approach, and similar shifts can be witnessed in other curriculum settings. However, the reality of pendula is that they keep on swinging and, in the context of educational assessment, often create unhelpful shifts and tensions in policy that have challenging impacts in classrooms for both teachers and learners.

The pendulum swing in the context the English NC has been a turbulent tug-of-war between the educationalists and politicians, policy makers and industrial stakeholders. The former have brought a focus on the learner to centre stage while the latter groups have brought priorities such as the economy to the forefront. Technology Education can be caught in the crossfire of such battles as, for example, has been seen in England where the GCSE (16+ exit examination) has seen the balance between continuous assessment of coursework (which suits the practice-based ethos of D&T) and end of course summative assessments swing dramatically. From 60 % coursework when GCSEs were introduced in 1988, the first shift was in 2009 to a maximum 60 % 'controlled assessment' (i.e., coursework conducted under exam conditions so that there could be no 'cheating' or 'parental help') to a 2013 proposal to remove all coursework from GCSEs, with the exception of practical subjects (including D&T) where an amount determined by strict principles may be allowed (Ofqual 2013).

## Contested and Conflicting Philosophies

Underpinning conflicting positions of different stakeholders is the age-old dichotomy of what we are educating learners for. Where school education is seen to be important in the context of the rounded education of the whole individual for life, there is typically an emphasis on intrinsic drivers for learning and, following from this, attainment and achievement. Where a more instrumental view is taken, assessment tends to be more focused on extrinsic drivers of attaining the right grades and, particularly in exit assessments, towards fitting into the workplace of a society.

When the economic position of a society is under threat, policy makers turn to education as a way of solving this problem. This can be seen throughout history as industrialisation and—more recently globalisation—have made their mark, the result of which has been to politicise assessment and increase the emphasis on more instrumental goals. This shift can be seen in the increased attention that governments pay not just to national comparisons of achievement but to where they stand on the international stage as identified through assessments such as the PISA tests. The emphasis on instrumentally-focused assessment encourages what might be seen as the more traditional positivist approach to assessment based on predetermined outcomes and performance criteria that "takes away from the originality, criticality and creativity of the work" (Elton 2006, p. 124).

Once assessment becomes politically loaded, systems are thrown into tension between the focus on the individual or on whole populations and on local agendas or global agendas. While educators are more likely to favour individual approaches that focus on culturally and socially relevant assessment, politicians concerned with their rankings in international league tables are more likely to focus on generalised and traditional knowledge-focused assessments that match the positivist views described above by Elton.

Dipping into examples highlights the challenges that these tensions bring. Pellegrino (2006), analysing the problems of a "flawed and broken" assessment in the USA, sees a system in need of radical change

> ... so that it can support processes of teaching and learning focused on deep learning and understanding ... the dollars we now spend on an assessment should be reinvested in more targeted and efficacious assessment approaches tied to important curricular goals. These assessments should be meaningful to the individuals assessed and have real value in determining their readiness to move on in the educational system. (p. 2)

Working alongside other educationalists and also industrialists, Pellegrino contributed to the National Centre on Education and the Economy's *New Commission on the Skills of the American Workforce* (NCEE 2007) to seek a solution to this 'broken' system. The report, laying out the territory for a new approach, makes it clear that the Commission's concern is with the steep decline in the United States' place in international league tables of educational attainment. Reflecting back to the report by the first Commission in 1990, the authors comment that

> [t]he first Commission never dreamed that we would end up competing with countries that could offer large numbers of highly educated workers willing to work for low wages. But China and India are doing exactly that. Indeed, it turns out that China and India are only the tip of the iceberg. . . . Thirty years ago, the United States could lay claim to having 30 % of the world's population of college students. Today that proportion has fallen to 14 % and is continuing to fall. (p. 4)

Where assessment is concerned, the report calls for development of "standards, assessments, and curriculum that reflect today's needs and tomorrow's requirements" (p. 14) and criticises current school exit exams that "measure the acquisition of discipline-based knowledge in the core subjects" (p. 14), wishing to see these replaced with qualities they perceive to be needed in the twenty-first century, such as creativity and innovation, ideas and abstractions, self-discipline, functioning in a team, etc. This report has led to the recent development of "Excellence for all", a pilot programme that has created an alternative curriculum and assessment system for high school students designed to lead them to higher education or skilled employment. The system focuses on the new 'Core Curriculum', which excludes Technology Education.

The issue of vocationalism is also a very real one for Technology Educators—where the balance between educating for life and educating for a technological job can bring different priorities both in curriculum and in assessment, resulting in summative assessments that focus less on a holistic view of capability within a practice-based discipline and more on specific and isolated vocationally-related knowledge and skills. This split view is compounded by different paradigms operating within curriculum and assessment—what Shepard (2000) identifies as a disjuncture between assessment systems still operating on a behaviourist paradigm while curriculum and 'instruction' have moved towards new paradigms of constructivist and sociocultural learning (see Chap. 5 by David Mioduser and Chap. 6 by Weny Fox-Turnbull, this volume).

These alternative views of learning can be seen as underpinning some of the contradictions—that even when teachers have progressive approaches to learning and teaching, they still operate within a non-aligned model of traditional assessment. Furthermore, even when policy has moved forward, teachers may still cling to traditional practices. Of course, this issue is not unique to the USA. For example, it has been noted by Beets and van Louw (2011) in the context of policy developments in South Africa, where they found that policy is ahead of many teachers whose teacher-directed approaches create an ethos that suggests assessment is something that teaches do to learners, not that learners are engaged in for themselves.

## The Impact of Assessment Systems on Learners and Learning

Few people enjoy being tested and for some the level of stress is entirely counter-productive in relation to learning. This is exacerbated when assessment is divorced from the processes of learning. Harlen and Deakin Crick (2002) explored the impact of high-stakes assessment policy on learner motivation. Analysing existing research from a range of countries (Canada, Israel, Morocco, Northern Ireland, UK and USA) on the impact of summative assessment on motivation, they found that where there was impact, it was largely negative. They reported that low achieving learners experienced a lowering of self esteem when faced with summative assessment, that learners (and especially girls) suffered anxiety, and that the ethos in classrooms changed such that learners perceived all assessments as summative even when the teachers intended them as formative. Teaching styles became more transmissional (echoing the disjuncture of paradigms highlighted by Shepard 2000), which created a bias towards learners who responded well to this style. The emphasis on learning shifted from process to performance and students felt the pressure of high stakes assessment whether the consequences were personal (as in the eleven plus exams in Northern Ireland) or more school focused (as in England's NC assessments or the SAT assessments in the United States).

Harlen and Deakin Crick (2002) drew important messages for both practice and policy from their study, suggesting a shift in practice to focus on process over performance, to develop a constructive and supportive ethos to assessment, to cultivate intrinsic motivation and self assessment, and for policy makers to additionally consider the issues the study raised about the validity of assessments that had such negative and cumulative impacts on learners. This research indicates important issues for Technology Education. As a process-based learning area where creativity and risk-taking are vital, learning needs to be enacted in environments that nurture self-esteem and motivation if learners are to develop confidence and competence as technologists and designers.

A further set of issues, focusing on equity, are presented by Beets and van Louw (2011) through an analysis of the impact of policy on assessment practices in the context of post-Apartheid South Africa. With reference to Bourdieu's concept of cultural capital, they draw attention to the extent to which assessments are influenced by the way in which educational resources are distributed and the fact that assessments are aimed at a societal norm:

> aimed at able-bodied learners, who are in command of the (instructional) languages of the assessors, who are appropriately prepared (had access to good teaching, etc.) and who are able to read, write and understand what is presented to them. (p. 311)

For South Africa, as with many countries where multiple indigenous languages sit alongside a mainstream language, the language of assessment is critical in ensuring validity and reliability. In the South African context, learning and assessments are carried out for the majority of children in a second or even a third language. Beets and van Louw draw attention to the extent to which the learner's own language, a key aspect of their cultural capital, is being marginalised through the medium of assessment, with consequent impediment on individuals' ability to achieve.

Each of the examples above illustrates the tensions that arise between the assessment policies that governments create and the practices that are enacted in classrooms. Each also draws attention to the impacts on learning and teaching that these tensions create. What is abundantly clear is that any assessment system is likely to have unintended consequences. But the more these consequences are brought to the fore, the greater the chance that any new systems can take into account and minimise the untoward.

## Assessment Structures and Practices

While it is useful to understand the drive behind instrumental views of assessment, educators are more likely to see value in approaches that are designed to support the development of the learner and by inference the learning experiences that teachers provide. Viewed from the standpoints of reliability and validity, positivist, instrumental approaches are likely to be more concerned with reliability and interpretivist, liberal approaches more concerned with validity. As government agendas are often the dominant driver in educational assessment, educationists are often put in the position of seeking the 'added value' of validity to get the best out of a system on behalf of the learners. This position could be seen as one of compromise and I would argue that we should (always) be seeking to optimise a system. I would also argue that future-facing assessment structures and practices in Technology Education should prioritise validity concerns and then aim to gain added value from achieving reliability.

This view accords well with the concept of sustainable assessment developed by David Boud (2000; Boud and Falchikov 2007) through which, in reference to Higher Education assessment practices, he proposes

> assessment practices that met the needs of an institution to certify or provide feedback on students' work, but which would also meet the long-term need of equipping students for a lifetime of learning (Boud 2000). In this way of viewing assessment, every act of assessment would do double duty in addressing the immediate needs of the institution while also contributing in some way to the development of the skills and capabilities of students to be effective learners. (Boud and Falchikov 2007, p. 7)

Anning et al. (2009), working from a sociocultural-historical perspective in Early Childhood Education, make a similar point:

> In working towards the future, early childhood teachers need two types of conceptual tool. The first tool is built upon socio-cultural theory, where documentation of learning moves beyond an individualistic orientation and acknowledges that learning is owned by a community of learners. In building learning stories and in mapping the transformation of understanding greater insights can be gained about children's learning and teachers' teaching. Secondly, the profession needs instruments which can extract from this rich web of assessment activity discrete measures of understanding as matched to government priorities. (p. 194)

Identifying an aspiration to achieve this optimization in assessment is one thing. Finding the practices that support the aspiration is something else. Critical in doing this is, I believe, the need to identify some underlying principles that can help all stakeholders, and most importantly teachers, maintain a balance between the different drivers of assessment policies and practices and take ownership of how these impact on learners and learning.

The issues raised so far in this chapter, including the need to shift the dominant behaviourist assessment paradigm to align with more social constructivist learning provide some indications that can be drawn on in identifying such principles. Research in D&T education that we have conducted over nearly 30 years in the Technology Education Research Unit (TERU) complements and grounds the above and it is to this research that I now turn. On the basis of both the issues raised and on grounded research I propose the following are key in creating guiding principles:

- Adopting a pedagogic approach to assessment
- Maintaining authenticity in activities through which assessment is being undertaken
- Recognising the importance of judgement in valid processes of assessment
- Maintaining a focus on equity and the inclusive role of the learner.

## A Pedagogic Approach to Assessment

If we believe that assessment is fundamental in learning, then is there any learning activity that, potentially, can't be optimised by also being an assessment activity? Likewise, is there any assessment activity that can't be optimised in terms of how it also supports learning? While there are clearly plenty of examples of both teaching and assessment where this is not the case, in an ideal world why would we not want it to be so? Where teaching and assessment practices explicitly aim to support learning, we have what can be termed a pedagogical model of assessment. In many ways such a model can be seen as an efficient one, particularly in terms of the time saved from the preparation for, and taking of, more traditional assessments.

This approach was one that we took when creating Technology Standard Assessment Tasks (SATs) for 5–7 year olds at the introduction of a Technology NC in England and Wales in 1990. The approach was important. Not only were teachers not experienced in the criterion and levels-based assessment that the NC introduced, but it was the first time that Early Years teachers were teaching a subject formally called Technology. For many, teaching what they saw as a new subject was daunting enough, let alone having to assess the learners as well. Consequently we focused on the development of good learning and teaching activities designed to scaffold teachers' understanding of structuring and implementing Technology projects that would also provide evidence of learning. The activities were accompanied by exemplified guidance for assessment against the NC criteria (Stables 1992). Effectively we were supporting the development of the teachers' Pedagogic Content Knowledge (PCK), although this was not the

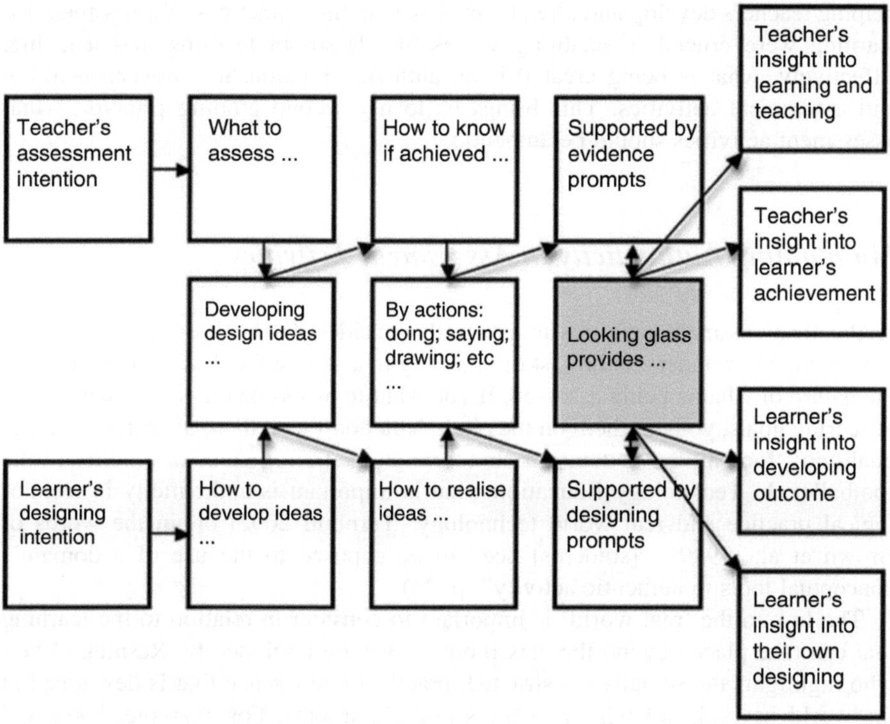


**Fig. 7.1** The mirror effect of effective evidence prompts (developed from Stables 1992 and Kimbell and Stables 2008)

language used at the time. The sets of SATs that we generated were designed to model learning, teaching and assessment in Early Years Technology activities. The approach was one that saw learning and assessment as symbiotic. The assessment activities were structured to work as stand-alone learning and teaching activities, and yet also as formative, diagnostic and summative assessment activities. In developing them we were mindful of the relationship between the learning intention, the assessment intention and the activity designed to support both—what Biggs (1996) terms "constructive alignment" (p. 347).

More recently we have likened this process to seeing this activity through a mirror where both teacher and learner can see double-sided reflections supporting summative and formative assessment and also learning and teaching (Kimbell and Stables 2008), as shown in Fig. 7.1.

The growth of Technology Education across the globe has put many teachers in the position of the Early Years teachers in England and Wales when the NC was introduced. The importance of supporting linked teaching and assessment—what I am calling a pedagogic model of assessment—is critical. Moreland and Cowie (2009) support this idea further through their research in the InSiTE project (Interactions in Science and Technology Education), where the importance of helping teachers develop and align their PCK with their practices of assessment for learning were crucial in enabling successful classroom learning and teaching. Effectively, what is being created is an authentic relationship between learning and assessment activities. This brings us to my second guiding principle—that assessment activities should be authentic.

## *Maintaining Authenticity in Assessment Activities*

Authentic assessment activities are generally considered to be those where a learner is showing their understanding, skill or ability in a setting that has some validity to the nature of what is being assessed. If you want to assess how good a footballer is at scoring goals, you put them on the pitch. You don't give them a written test. The quality of 'footballing' is demonstrated through the performance or practice of the footballer. In Technology Education what is important is authenticity in technological practice and real world technology (Turnbull 2002) or, in the words of Brown et al. (1989), "[students] need to be exposed to the use of a domain's conceptual tools in authentic activity" (p. 34)

The link to the 'real world' is important to consider in relation to the learning that can take place beyond the classroom, as we are reminded by Resnick (1987) who highlights the socially constructed, practical intelligence that is developed in the world beyond, not within, schools and classrooms. For example, Fleer and Quiñones (2009), drawing on a study with primary aged children, highlight the funds of technological knowledge that learners both bring and gain through the informal setting of a school 'tinkering club' that capitalised on learners' interests in the materiality of technological artefacts and provided an experiential and social setting in which to build understanding from these interests. Both examples highlight the reality that, removed from formal classroom settings, informal authentic activities often support learning in ways that are different to formal classrooms, through practical, social, physical (hands-on) and concrete (rather than abstract) situations. The challenge is to bring the real-world authenticity of such activities into classroom-based learning and assessment. Technology Education offers real opportunities here—the 'real-worldness' of technological practice in classrooms can take learners beyond the realms of learning abstract knowledge and skills and into the social and cultural settings in which the practice takes place while also taking into consideration the human needs and wants that are driving the intentions behind the practice. This allows for further authenticity by engaging learners in assessment tasks that are embedded in contexts that are relevant and motivating to learners, supporting their taking ownership of their learning (see also Chap. 6, this volume, by Wendy Fox-Turnbull).

Learners need plenty to get their teeth into and be challenged by. Engaging them in relevant, issues-rich tasks is a good way of allowing them to both develop and demonstrate their capability. But creating such contexts is not straightforward. There is a history of Technology Education learning tasks being teacher-led and driven by the dominant culture of an education system, which may be quite different and socially and culturally irrelevant to the learners being assessed. Those devising assessment contexts can be well meaning in attempts to create motivating contexts but may be working on assumptions about what will inspire learners to show what they can do. This was highlighted in a recent study concerning the contexts in which learners in Malawi expressed interests in the context of learning mathematics (Kazima 2013). The study identified social relevance as

> that which connects with the present and future lives of students as well as the issues that are of importance to them and their communities, and in the interest of humankind in general (p. 23)

and cultural relevance, as defined by Ladson-Billings (1994), as education

> that empowers students intellectually, socially, emotionally and politically by using cultural referents to impart knowledge, skills and attitudes. (pp. 17–18)

Kazima highlights the lack of voice of the learners, commenting that:

> Issues of relevance are often decided by a number of groups of people including policy makers, curriculum developers, textbook writers, and class teachers. Policy makers are often guided by the issues affecting the nation at the time. … Curriculum developers interpret the policies and develop the curriculum as they see 'relevant' to the nation's needs. Textbook writers interpret the curriculum into a possible form of teaching. Finally teachers interpret both the curriculum and the textbooks into lessons. (p. 25)

Using a survey tool previously used in an international study on relevance in mathematics teaching (Julie and Holtman 2008) and customised for Malawi, Kazima goes on to show that learners who were surveyed rated contexts relating to modern technologies more highly than contexts relating to agriculture and what are termed ethnomathematics (e.g., the mathematics used in basket weaving). With hindsight, these findings can be explained in terms of learners' curiosity and aspirations, but given Malawi's rural context, it is clear to see how a potential mismatch can arise between the issues of relevance likely to be identified on behalf of the learners by policy makers, etc., and the very things that would inspire learners to see mathematics learning as relevant.

In piloting a parallel survey, customised to the culture of England and D&T lessons, findings had some fascinating parallels and also differences (Stables 2013). Both groups of teenagers showed similar levels of interest in new technologies addressing issues of health and of secrecy and privacy. Both also showed a tendency to reject certain views of the policy makers—the Malawi maths learners rejecting agriculture and ethnomathematics, while the English D&T learners rejected learning about the lives of famous designers (popular in examination syllabi) and horticulture, a context that a former Minister for Education was keen to see adopted in the 2014 revision of the English D&T National Curriculum. The lesson to be learned is that we need to focus more on how the learners themselves are engaged with selecting and building the contextual backgrounds for their tasks. This is not to deny the contribution of the teacher, but to suggest how the choice and development of authentic contexts can be optimised.

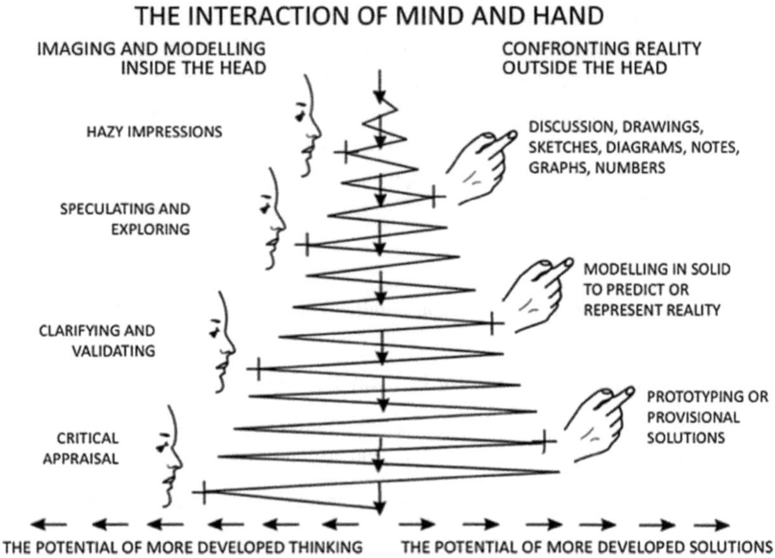


**Fig. 7.2** The APU Design and Technology Model (Kimbell et al. 1991, p. 20)

However well considered the context of an activity is, in technological practice bringing an authentic view of process is of equal importance (Moreland 2009). When the research team in TERU was commissioned in 1985 by the UK Department for Education's Assessment of Performance Unit (APU) to assess the D&T capability of 10,000 15 year olds, one of our first challenges was to understand the design processes that capability was evidenced through. Full accounts of our encounter with, and conclusion to, this challenge can be found elsewhere (Kimbell et al. 1991; Kimbell and Stables 2008). Briefly, we came to the challenge with a distrust of the linear and cyclical models of process that were common in the literature and assessment practices at that time, because of the lack of authenticity they demonstrated. With their prescriptive and managerial structures they denied both the reality of the complex and diverse ways in which ideas are generated and the processes through which ideas grow to become thoughtful and well-developed working realities. Drawing on our experience as designers and teachers, alongside the performance data collected through our research, we created a different model where the line of development was from hazy to clear and the process was driven by the iterative interaction of action and reflection (see Fig. 7.2). This model allowed us to understand the various ways in which the 10,000 learners in question went about their processes of designing and (along with a team of 100 teacher assessors) come to conclusions about the learners' levels of performance in D&T.

While over the years we have come to have richer understandings of the ways individuals approach processes of designing, not least through others exploring similar territory (e.g. Buchanan 1995; Darke 1979; Lawson 1990, 2004; Middleton 2005; Nelson and Stolterman 2003), we have continued to find the model useful in understanding the reality of the complex processes at work. This same model of process continues to underpin our work on assessment, including our current work on digital assessment portfolios. While the model is challenging from a manageability point of view, assessment that respects the authenticity of individual processes is important in considering authenticity from a learner's perspective. The route one learner takes through the process will be different to another. A teacher accepting the iterative and responsive nature of the model needs alternative strategies to manage the assessment and, by implication, the learning. Our experience from research suggests that the pivot points in the activity are action and reflection: supporting learners to explore their thinking through action and then pausing to reflect on that action is a fundamental rhythm to encourage.

The final piece in this jigsaw of authenticity is of the criteria against which assessment judgements are made. This takes us back to the example given earlier of assessing a footballer's ability to score goals. The criteria need to be 'authentic' in terms of what they are attempting to reveal and they need to be applied in a situation where they can be validly evidenced. "Can score goals" is unlikely to be evidenced fully or validly in a written test.

Reflecting back to the time when we created the APU D&T model of process, our hunch was that criteria would be derived from key elements of the process (identifying and addressing issues in the task; having a grip on generating ideas and developing solutions; appraising thinking with a sound, critical eye) and interconnectedness of the iteration between thought and action. Moreover, with a belief that the value of a whole enterprise is greater than the sum of its parts, our 'hunch' was to view all evidence holistically. Taking account of the whole of what a learner has set out to do, how they have gone about it, and what they have achieved, enabled us to see elements of process (evaluating, generating, researching, problem solving, etc.) at whatever stage they appeared and to take an overall position on the learner's achievements. Looking within the evidence allowed us to diagnose other aspects—strengths and weaknesses of their approach, how they had understand the application of knowledge, etc. This approach may seem counter-intuitive but, for the record, the most statistically reliable judgements the teachers (assessors) made were the holistic ones. The smaller and more atomised the assessment decisions became, the less statistically reliable they were found to be (Kimbell et al. 1991).

## *The Importance of Judgement*

Our research has consistently indicated that engaging in holistic assessment acts as an important professional development tool for teachers and an important learning tool for learners. A facet of this is the emphasis on making judgements of qualities, as opposed to awarding marks based on 'right' or 'wrong' answers. Boud (2007) suggests that judgement is "the capacity to evaluate evidence, appraise situations and circumstances astutely, to draw sound conclusions and act in accordance with this analysis" (p. 19).

Making a judgement is a complex process, but that does not necessarily mean that it is difficult or unreliable. Judgements are not made in a vacuum and, in educational contexts, often involve drawing on an individual's repertoire of professional experience. It has been our experience that when a teacher or learner is asked to make an overarching judgement about what has been achieved in an assessment task this allows them to reflect carefully on the evidence and the context in which it has been created. If asked to share and debate their judgements, whether in assessment moderation meetings or in self and peer assessment activities, thinking becomes clearer as participants articulate and justify their positions. It is often in this dialogue that teachers and learners grow in both their understanding and confidence.

Boud (2007) identifies the act of making a judgment as moving an individual from passive recipient to active assessor. Where learners are concerned he sees a double value—learning to form judgements as well as learning to act on the judgements formed. Drawing on both social theory (Bourdieu and Wacquant 1992; Giddens 1991) and psychology (Karoly 1993) he goes further to propose that bringing judgement to the fore in assessment helps develop both reflexivity and self-regulation in learners as they enable individuals

> to 'look again', to monitor one's own performance, to see one's own learning in the context in which it is deployed and to respond with awareness to the exigencies of the tasks in which one is engaged. Reflexivity and self-regulation . . . involve dispositions and an orientation to both work and learning. They also have an affective dimension. They involve confidence and an image of oneself as an active learner, not one solely directed by others. A focus on reflexivity and self-regulation is a key element in constructing active learners, as these features need to be constructed by both teachers and examiners themselves. (pp. 21–22)

The approach to using judgement that we have taken in recent research (the e-scape project) has been based the law of comparative judgment (Thurstone 1927), which, at its simplest level, is based on detailed, holistic reviewing of two different pieces of work and then making a judgement about which is better. Linked to a computer-driven algorithm that presents 'judges' with series of pairs of work to be judged, a rank is created. Our development and use of this process is described in detail elsewhere (e.g., Kimbell et al. 2009). As an assessment process, statistically it has very, very high reliability. But for our purposes here, it also presents potential for professional development of teachers and metacognitive understanding for learners, as a concrete example of the reflexivity and self-regulation identified by Boud, referred to above.

In a pilot project, 15 year olds who had taken part in the e-scape assessment activity then became part of the judging team. Not only were their judgements consistent with the adults, they found the exercise highly illuminating because of the insights gained from assessing each other's work. They commented, for example, that they felt better prepared for future work (Kimbell 2012). In summing up their experience, their teacher (who had also engaged in the research as a judge) commented

> It was unbelievable how quickly all of the students managed to get to grips with the assessment process. And listening in to the conversations that were going on during the judgments it was apparent that they didn't have many of the hang-ups that I as a teacher have experienced. Without a doubt they were able to spot the creative and innovative thinking in the design work and were rarely taken in by "pretty" or "content free" products. (Kimbell et al. 2009, p. 161)

Not only does this example illustrate the value of engaging learners in holistic judging as a form of peer and self assessment, it also indicates new possibilities in democratic approach in the context of high stakes assessment, this pilot being part of a trial for authentic assessment that could be used in high-stakes contexts.

## *The Importance of Equity*

As an example of democratic assessment, the previous example of e-scape places the assessed in the active role of the assessor. But not all assessment systems are so inclusive. To be equitable, assessment practices need to ensure the assessment process itself is fair, for example, taking account of the ways different learners might demonstrate achievement and attainment, including through valid but less tangible modes, such as group work and talk. Referring back to Shepard's (2000) concern that assessment practices have not progressed in step with understanding of effective ways of learning, particularly through sociocultural approaches, there is a real danger that assessment is inherently unfair, and therefore invalid, if misaligned with the real evidence of learning.

Ensuring that learners have the opportunity to demonstrate their achievements in appropriate ways requires shifts in thinking around the nature of evidence. For learners with special educational needs it may be something of a worthless task to attempt to assess their understanding through a written test if this is not the best way for them to communicate. For learners whose first language is not that of the dominant educational culture, then assessing them through this language is unlikely to provide a true reflection of what they can achieve. Both of these examples come from what might be seen as 'special' situations. But given the nature of Technology Education and the ways in which technological literacy and capability can be enacted, and given the understandings we now have about preferred learning styles and even designing styles (Lawler 1999, 2006) it seems not just iniquitous but also inefficient not to take these into account when considering effective assessment approaches. With the increased possibilities of digital tools for capturing data in diverse ways there is a real opportunity to use these tools to support learners to communicate the evidence of their technological practice in ways that genuinely and appropriately demonstrate their capability.

## The Impact of Digital Technologies on Assessment

Digital tools can be used effectively to support assessment of different learning and designing styles. But just because a tool is digital, its use in assessment doesn't, by definition, make the approach better. If we use digital tools to do things that were wrong in an analogue world, they won't become right just because of the use of 'new' technologies. However, when considering the affordances of digital technologies, including in relation to the underpinning principles that have been outlined above, it can be seen that they offer positive benefits to various aspects of assessment processes and practices.

An immediate benefit is the way evidence can be organised in digital structures, an approach used extensively in e-portfolios. Presenting both the process and the outcome of technological activity digitally allows a rich collection of text, image, audio and video to be included. E-portfolios are increasingly commonplace, even for high-stakes assessment. Pragmatically they allow for the submission, storage and archiving of assessment evidence in an accessible way that isn't physical space hungry. Creating a summative assessment portfolio can be a useful activity in itself as the learner reviews and selects work to submit. But this approach can fall into the same trap as when it is done through pencil and paper if valuable learning time is wasted in re-presenting work for the assessor, denying the real benefits of dynamic digital capture using digital tools to 'hoover up' the evidence as it is generated. This latter approach draws on the range of tools that are available for documenting written and spoken word alongside image-based data.

The mobility of digital tools, with smart phones, netbooks and laptops, also can contribute to effective and authentic approaches to assessments since the collection of digital evidence can be undertaken in a range of settings, rather than being tied to regular classroom activity or examination halls. For example, assessments that are in the workplace setting can readily collect evidence of the performance of activities in the workplace. This benefit can also be seen for activities that are best captured through their physical enactment where learners are demonstrating their understanding 'for real' either through practical activities or the presentation of their ideas.

As has earlier been outlined, there are benefits of using the practice of holistic judgement in assessing project work. When linked to web-based systems, digital tools can make this process more manageable, allowing multiple assessments to take place at any given time and in multiple locations (Kimbell et al. 2009). Web-based digital portfolios can also be hugely beneficial for formative assessment, as we found when trialing the e-scape system in a number of schools in Israel (Stables and Lawler 2011). Learners felt that their teachers had a better understanding of the process they had gone through in their projects and teachers felt that it benefitted a broader range of learners, even the special needs children finding assessment more supportive. Seery et al. (2012) have also shown how the approach proved both valid and reliable in undergraduate peer assessment of design project work.

Each of the above dimensions of using digital tools in assessment has the added value of supporting flexible and agile learning. In the context of the dynamic nature of technological projects, digital tools have the dual advantage of supporting the learning itself while also documenting the evidence as it is created. Digital tools can play an important role in supporting future approaches to assessment in Technology Education by adding value to existing effective approaches and by the new functionalities that they bring to help address age-old problems such as reliability of judgements in performance settings.

## The Future Agenda

Through this chapter a range of issues, ideas and insights have been presented. History has shown the damage done in the name of assessment but research has provided some clarity and grounding that provides direction to positive avenues for moving forward.

It is particularly useful to consider how the paradigm that dominates assessment practices still largely conflicts with more progressive understandings in learning and teaching. This gives a clear message about the importance of leading assessment developments through the lenses of more progressive paradigms of learning and teaching and helping all stakeholders, and especially teachers, to understand both the theoretical and practical underpinnings to move Technology Education assessment to more constructivist and sociocultural approaches. It is interesting to note how, in Technology Education research, developments in adopting sociocultural approaches to assessment are largely in the field of early years and primary education. Much has to be gained from extending this research to the secondary years, which have been more dominated by high-stakes summative assessment. Placing a greater emphasis on social and cultural relevance in assessment activities will also open up new understandings to support more equitable approaches.

The chapter has also indicated the potential affordances of new technologies in assessment and this is an area that will inevitably play an increasing role in education systems. An important message here is to ensure that digital approaches build on and develop existing sound and authentic approaches such that these are enhanced. This links to the overarching message of the chapter—to continue to push for assessment that is authentic at all levels. The practice-based, real-world nature of Technology Education provides an excellent, sympathetic setting for the development of authentic approaches to assessment. Core to future developments should be continued support for teachers to align learning and assessment in order to engage and motivate learners through technological practice that makes visible to the learners and their teachers and assessors the learning that has taken place and the capability that has been developed.

## References

Anning, A., Cullen, J., & Fleer, M. (2009). *Early childhood education: Society and culture* (2nd ed.). London/Thousand Oaks/New Delhi/Singapore: Sage.

Atkinson, S. (2000). Does the need for high levels of performance curtail the development of creativity in design and technology project work? *International Journal of Technology and Design Education, 10*(3), 255–281.

Beets, P., & van Louw, T. (2011). Social justice implications of South African school assessment practices. *Africa Education Review, 8*(2), 302–317.

Biggs, J. (1996). Enhancing teaching through constructive alignment. *Higher Education, 32*, 347–364.

Boud, D. (2000). Sustainable assessment: Rethinking assessment for the learning society. *Studies in Continuing Education, 22*(2), 151–167.

Boud, D. (2007). Reframing assessment as if learning were important. In D. Boud & N. Falchikov (Eds.), *Rethinking assessment in higher education: Learning for the longer term* (pp. 14–26). Abingdon: Routledge.

Boud, D., & Falchikov, N. (2007). Introduction: Assessment for the longer term. In D. Boud & N. Falchikov (Eds.), *Rethinking assessment in higher education: Learning for the longer term* (pp. 3–13). Abingdon: Routledge.

Bourdieu, P., & Wacquant, L. (1992). *An invitation to reflexive sociology*. Chicago: University of Chicago.

Brown, J. S., Collins, A., & Duguid, P. (1989). Situated cognition and the culture of learning. *Educational Researcher, 18*(1), 32–42.

Buchanan, R. (1995). Wicked problems in design thinking. In V. Margolin & R. Buchanan (Eds.), *The idea of design* (pp. 3–20). Cambridge, MA: MIT.

Darke, J. (1979). The primary generator and the design process. *Design Studies, 1*(1), 36–44.

DES/WO. (1988). *National Curriculum: Task group on assessment and testing. A report*. London: Department of Education and Science and the Welsh Office.

Elton, L. (2006). Assessing creativity in an unhelpful climate. *Art, Design & Communication in Higher Education, 5*(2), 119–130.

Fleer, M., & Quiñones, G. (2009). Assessment of children's technological funds of knowledge as embedded community practices. In A. Jones & M. de Vries (Eds.), *International handbook of research and development in technology education* (pp. 477–492). Rotterdam: Sense.

Fleming, R. (2013). *Design education for a sustainable future*. London/New York: Earthscan (Routledge).

Giddens, A. (1991). *Modernity and self-identity: Self and society in the late modern age*. Cambridge: Polity Press.

Hanson, F. A. (1994). *Testing, testing: Social consequences of the examined life*. Berkley/Los Angeles: University of California.

Harlen, W., & Deakin Crick, R. (2002). A systematic review of the impact of summative assessment and tests on students' motivation for learning (EPPI-Centre Review, version 1.1). In *Research Evidence in Education Library* (Vol. Issue 1). London: EPPI-Centre, Social Science Research Unit, Institute of Education.

Harris, M., & Wilson, V. (2003). *Designs on the curriculum? A review of the literature on the impact of Design and Technology in schools in England*. Nottingham: Department for Education and Skills.

Julie, C., & Holtman, L. (2008). The Relevance of School Mathematics Education (ROSME). In L. Holtman, C. Julie, O. Mikalsen, D. Mtetwa, & M. Ogunniyi (Eds.), *Some developments in research in science and mathematics in Sub-Saharan Africa: Access, relevance, learning and curriculum* (pp. 379–405). Somerset West: African Minds.

Karoly, P. (1993). Mechanisms of self-regulation: A systems view. *Annual Review of Psychology, 44*, 23–52.

Kazima, M. (2013, January). *Relevance and school mathematics*. Paper presented at SAARMSTE 2013: Making mathematics, science and technology education socially and culturally relevant in Africa, University of the Western Cape, Cape Town.

Kelly, A. V. (2009). *The curriculum: Theory and practice* (6th ed.). Los Angeles/London/New Delhi/Singapore/Washington: Sage.

Kimbell, R. (1997). *Assessing technology: International trends in curriculum and assessment*. Buckingham: Open University.

Kimbell, R. (2006). Innovative performance and virtual portfolios—a tale of two projects. *Design and Technology Education: An International Journal, 11*(1), 18–30.

Kimbell, R. (2012). Evolving project e-scape for national assessment. *International Journal of Technology and Design Education, 22*(2), 135–155.

Kimbell, R., & Stables, K. (2008). *Researching design learning: Issues and findings from two decades of research and development*. Berlin: Springer.

Kimbell, R., Stables, K., Wheeler, T., Wozniak, A., & Kelly, A. V. (1991). *The assessment of performance in design and technology*. London: SEAC/HMSO.

Kimbell, R., Wheeler, T., Stables, K., Shepard, T., Martin, F., Davies, D., et al. (2009). *e-Scape portfolio assessment*. A research & development project for the Department of Children, Families and Schools, phase 3 report (p. 169). London: Goldsmiths, University of London.

Ladson-Billings, G. (1994). *The dreamkeepers: Successful teachers for African-American children*. San Francisco: Jossey-Bass.

Lawler, T. (1999). Exposing the gender effects of design and technology project work by comparing strategies for presenting and managing pupils' work. In P. H. Roberts & E. W. L. Norman (Eds.), *IDATER 99: International conference on design and technology educational research and curriculum development* (pp. 130–137). Loughborough: Loughborough University of Technology.

Lawler, T. (2006, December). *Design styles and teaching styles: A longitudinal study of pupils' ways of doing designing following complementary re-grouping and teaching*. Paper presented at TERC 2006: Values in Technology Education, Gold Coast, Australia.

Lawson, B. (1990). *How designers think: The design process demystified* (2nd ed.). Oxford: Butterworth Architecture.

Lawson, B. (2004). *What designers know*. Oxford: Elsevier.

MacLeod, R. M. (1982). *Days of judgement: Science, examinations and the organization of knowledge in late Victorian England*. Driffield: Nafferton Books.

Middleton, H. (2005). Creative thinking, values and design in Technology Education. *International Journal of Technology and Design Education, 15*(1), 61–71.

Moreland, J. (2009). Assessment: Focusing on the learner and the subject. In A. Jones & M. de Vries (Eds.), *International handbook of research and development in technology education* (pp. 445–448). Rotterdam: Sense.

Moreland, J., & Cowie, B. (2009). Making meaning in primary technology classrooms through assessment for learning. In A. Jones & M. de Vries (Eds.), *International handbook of research and development in technology education* (pp. 461–476). Rotterdam: Sense.

NCEE. (2007). *Tough choices or tough times: The report of the new Commission on the Skills of the American Workforce (Executive Summary)*. Washington: National Center on Education and the Economy.

Nelson, H. G., & Stolterman, E. (2003). *The design way: Intentional change in an unpredictable world*. Englewood Cliffs: Educational Technology.

Ofqual. (2013). *Review of controlled assessment in GCSEs*. London: Author.

Ofsted. (2001). *Ofsted subject reports 1999–00: Secondary Design and Technology*. London: Department for Education and Employment.

Ofsted. (2002). *Secondary subject reports 2000/01: Design and Technology*. London: Department for Education and Employment.

Pellegrino, J. W. (2006). *Rethinking and redesigning curriculum, instruction and assessment: What contemporary research and theory suggests*. A paper commissioned by the National

Center on Education and the Economy for the New Commission on the Skills of the American Workforce. Washington: National Center on Education and the Economy.

Prest, D. (2002). *An analysis of the attainment target level descriptors and associated programme of study in relation to the Design and Technology mission statement*. London: Department for Education and Skills, Design and Technology Strategy Group.

Resnick, L. B. (1987). Learning in school and out. *Educational Researcher, 16*(9), 13–54.

Seery, N., Canty, D., & Phelan, P. (2012). The validity and value of peer assessment using adaptive comparative judgement in design driven practical education. *International Journal of Technology and Design Education, 22*(2), 205–226.

Shepard, L. A. (2000). The role of assessment in a learning culture. *Educational Researcher, 29* (7), 4–14.

SSEC. (1960). *The Beloe Report: Secondary schools examinations other than the GCE*. London: HMSO.

Stables, K. (1992). The assessment of technology at key stage 1. In C. Gipps (Ed.), *Developing assessment for the National Curriculum* (pp. 42–50). London: Kogan Page.

Stables, K. (2007). Why so much emphasis on assessment? Editorial. *Design and Technology Education: An International Journal, 12*(2), 3–5.

Stables, K. (2013, December). *Social and cultural relevance in approaches to developing designerly well-being: The potential and challenges when learners call the shots in Design and Technology projects*. Paper presented at PATT: Technology Education for the future: A play on sustainability, Christchurch, New Zealand.

Stables, K., & Lawler, T. (2011). *Assessment in my palm: e-scape in Israel. Evaluation of phase 1*. London: TERU, Goldsmiths, University of London.

Taunton. (1868). *The Taunton report*. London: Schools Inquiry Commission.

Thurstone, L. L. (1927). A law of comparative judgement. *Psychological Review, 34*, 273–286.

Turnbull, W. (2002). The place of authenticity in technology in the New Zealand curriculum. *International Journal of Technology and Design Education, 12*, 23–40.

# Chapter 8
# Developing a Technology Curriculum

**David Barlex**

This chapter proposes three procedural principles that can inform the development of a technology curriculum: being true to the nature of technology, developing a perspective on technology, and enabling technological capability. It then explores possible futures for technology education curricula using each of these principles as a lens. Drawing on these scenarios, the argument is made that using these principles will result in a curriculum that is a valid and worthwhile endeavour for all students, and that facilitates the introduction of new elements, enabling the curriculum to keep pace with changes outside of school. The chapter concludes by asserting that teachers can use the principles to devise, justify and implement programmes of study that are robust and can withstand scrutiny from those who might question the worth of technology education. In turn, policy makers can benefit from the informed discourse with an articulate and knowledgeable profession.

## Introduction

This chapter is composed of three main sections. The first section considers how curricula are conceptualised and come into being. This leads to the idea that an important way to devise a curriculum is to formulate a set of procedural principles which can be used to develop and implement the curriculum. The second section identifies and discusses three procedural principles that can be used to develop and implement a curriculum for technology education: being true to the nature of technology, developing a perspective on technology, and enabling technological capability. The third section discusses possible future technology curricula in the

D. Barlex (✉)
Brunel University, Uxbridge, Middlesex UB8 3PH, UK
e-mail: david.barlex@btinternet.com

 
PJ. Williams et al. (eds.), *The Future of Technology Education*, Contemporary Issues in Technology Education, DOI 10.1007/978-981-287-170-1_8

light of the three procedural principles and its relationships with other school subjects. Finally there is a short summary. Before beginning the first section it is useful to consider possible arguments for teaching technology as different justifications will lead to different technology curricula. I can make four different arguments for teaching technology.

## An Economic Argument

A steady supply of people who have studied technology is essential to maintain and develop the kind of society we value. Technology is central to the innovation on which our future economic success as a nation depends. Technology qualifications open doors to a wide range of careers.

## A Utility Argument

It is useful in everyday situations to be able to use technological thinking, which includes design skills and technical problem solving, in being able to address and solve practical problems.

## A Democratic Argument

Through the media, people encounter issues that involve technology. Some understanding of technology is needed to reach an informed view on such issues and engage in discussion and debate.

## A Cultural Argument

Technology is a major achievement of our culture, so everyone should be helped to appreciate it, in much the same way that we introduce them to literature, art and music.

The question for you, the reader, is which—if any—of these do you think provides the best justification for teaching technology and, if more than one, in what order of significance would you place them? The answer you give may depend on whether you see technology education as something for ALL young people at school or whether you see it as something that applies to only a selected few.

The economic argument is difficult to justify as an argument for ALL young people as the total of professional engineers, technologists and designers is only a few per cent of the whole population of an industrialised country. Hence a primary goal of a general technology education cannot be to train this minority who will actually 'do' technology. The utility argument can be extended to a consideration of the personal qualities that being able to deploy design skills and technical problem solving develops. The creative activities of design and making—a major part of technology education courses—not only give immense personal satisfaction but importantly develop a sense of self efficacy which provides young people with a positive self image with regard to their ability to be successful. The cultural argument forces us to ask, "What are the grand narratives of technology that we want our young people to appreciate?" Who are the heroes of technology? Science has Newton, Galileo, Darwin, Einstein, Crick and Watson, Franklin, Higgs. In England science also has some high profile popularisers such as Brian Cox and Jim Al-Khalili. A problem facing technology here is that many manifestations of technology are the product of large multidisciplinary teams and the concept of individual 'heroes' does not apply. If we are to deploy a cultural argument then the technology curriculum will need to find the means within itself to do this. The role of education to produce informed citizens able to take part in rationale debate lies at the heart of the democratic argument. This is at the forefront of the recent report "Emerging biotechnologies: technology, choice and public good" (Nuffield Council on Bioethics 2013). It shows that such technologies are likely to have significant impact on society, almost certainly being disruptive of many current practices. Interestingly the report pays particular heed to the need for a public discourse about the development and deployment of such technologies. This is the nub of the democratic argument—enabling the public to contribute significantly and intelligently to any such discourse.

It seems to me that each of these arguments should inform a school technology curriculum. Although in particular circumstances their relative significance may vary, to produce a curriculum that did not respond in part to each of these arguments would be a curriculum that was lacking an important dimension.

## How Does a Curriculum Come into Being?

A curriculum has to be conceived and the principles on which this conception is based will to a large extent govern its nature and purpose. Often it is politicians and civil servants who take on this task, usually in collaboration with those outside government who have appropriate expertise. This was the case in England in 2011 when the Minister of Education, Michael Gove, appointed an expert panel to decide on a new National Curriculum. The brief for the expert panel was:

> To advise and make recommendations to the Department on the essential knowledge (e.g. facts, concepts, principles and fundamental operations) that children need to be taught in order to progress and develop their understanding in English, mathematics, science, physical education and any other subjects which it is decided should be part of the National Curriculum. (Department for Education 2012)

For some, particularly politicians, this seems relatively straight forward and it would appear at first glance that the important initial step is, as required by the brief, to define the knowledge that resides in the subjects to be taught. But this is more problematic than it seems, as the nature of knowledge is itself contested. There are some who hold an absolutist view of knowledge—for example, Immanuel Kant in the past and Richard Peters (1973) and Anthony O'Hear (1992) in more recent times—and for whom the knowledge exists 'out there' independent of the knowers. This has been criticised as perpetuating "a view of man [sic] as a dehumanised, passive object" with humans seen "not as a world producer but as world produced", leading to the assertion that such a view of knowledge is fundamentally dehumanising, ignoring the intentionality and expressivity of human action (Esland 1971, p. 70).

This view is confronting for the technology curriculum, where 'taking action' is seen as a fundamental aspect of technology. The absolutist position is challenged by the post modernist movement (e.g., Doll 1993) which completely rejects certainty and insists that what is known is dependent on the communities which have generated the knowledge. It is also subject to change as these and other communities contest that knowledge. For those wishing to put established, unchanging knowledge as a central organising tenet of the curriculum this is extremely uncomfortable.

But in tackling a technological task—and perhaps any task—there is the need for what might be termed 'reliable knowledge of the moment', that knowledge which a recognised community of practice has deemed fit for purpose in tackling the task. Hence while we as technology educators might acknowledge that technological knowledge is a socio-cultural construct, subject to change and interpretation, we will have to identify, describe and limit the knowledge we believe should be taught as part of the school technology curriculum. There is of course a tension here for those educators who adopt the position that in tackling a technological task the particular knowledge, skills and understanding that are needed to be successful should be acquired by the pupil on an as needed basis as part of tackling the task, as opposed to being explicitly taught as a precursor to tackling the task.

A further complexity for curriculum development is that politicians invariably want an education system in which it is possible to show that pupils are making progress. This has immediate implications. If the purpose of the curriculum is seen as the assimilation of the knowledge, however such knowledge is identified and selected, one way to monitor progress is to assess pupils with regard to their assimilation of the required knowledge. This has led to the production of documentation detailing a sequence of statements of attainment against which progress over time can be assessed—which, in turn, has led to the introduction of national formal testing arrangements in which pupil attainment is measured. Robin Alexander (1984) has questioned this in that "assessment via *formal* testing or other publicly verified procedures can only deal with very limited and specific areas of school work and is much less important than the continuous *informal*, (even unconscious) evaluation of the pupil by the teacher" (pp. 36–37).

The impact on classroom practice in England has in some cases been pernicious. Pupils are given level descriptors for particular requirements at the beginning of individual lessons and required to meet these through prescribed activities during the lesson. At the end of the lesson they discuss with the teacher and other pupils the extent to which they have met these requirements. In this way there is a continuous lesson-by-lesson monitoring of performance and both teachers and pupils are under pressure to ensure that progress is being made in terms of moving to higher levels of achievement as defined by the descriptors.

The problem here is twofold. The first is that the reason for learning the subject is lost and becomes reduced to achieving incremental, atomised features of learning which do not embrace the totality or worth of the subject. The second is the belief that pupils need to have clear and explicit criteria so that they can take responsibility for the orientation and the assessment of their own work. In principle this is a good idea but it is fraught with dangers. Enabling pupils to appreciate the meaning of learning criteria is not a trivial exercise. In addition, criteria to decide on the worth of a significant endeavour in technology will almost certainly involve holistic as opposed to atomistic judgment. The ability to make such judgement will be demanding of both teachers and pupils and needs to be built over time through discussion and reflection. In the light of this Paul Black (2013) commented to me, "The saying that appeals to me here is 'How do I know where I am going until I get there?' There's more truth in this than is evident at first sight."

While these issues are dealt with in more detail in Chap. 7 on assessment and Chap. 5 on pedagogy, here it is important to make two points. First, assessment of pupils in technology will inevitably be a feature of the curriculum and it is essential that curriculum be assessed in ways—both formative and summative—that are valid, reliable and adaptable to pupils' individual responses to the subject. Second, the use of assessment is a key feature of pedagogy that allows the teacher to gauge understanding and progress and as such should be considered explicitly as an integral part of pedagogy as opposed to an optional add on.

The process of devising a curriculum requiring the acquisition of previously identified knowledge and developing from this curriculum statements of attainment by which progress may be assessed has received significant criticism. It is not that such approaches produce curricula completely devoid of worth in various parts. Rather, it is that such curricula do not make explicit their own ideological position as to the overall purpose of their educational intentions (Kelly 2009). If an avowed purpose is for education to develop young people who can play a full part in a democratic society this requires a different perspective from the educator in order to ensure that curricula and practice in schools derived from curricula meet this requirement.

Stenhouse (1975) and Bruner & Haset (1987) amongst others have argued that to achieve this it is imperative to see curricula as process and development, and devise curricula accordingly. Hence it is important to identify procedural principles through which a curriculum may be devised, interpreted and implemented. This approach to curriculum takes cognisance not only of what the curriculum offers but also how it is offered. The important role of pedagogy in defining the 'how' of learning technology is discussed in detail by David Mioduser in Chap. in 5 and Wendy Fox-Turnbull in Chap. 6.

I warm to the idea of identifying a set of 'procedural principles' as advocated by Stenhouse and Bruner to inform a curriculum for two reasons. First it provides a framework which keeps sight of the nature of what is being taught and its worth. This is important as if teachers lose sight of this then the whole exercise ceases to be educational and is reduced to rituals of good test scores and qualification acquisition. Second it allows for interpretation by a teacher as to both the content of the curriculum and the way in which it is taught. This is important as the content of technology education is subject to change perhaps more than any other area of the curriculum and there is increasing choice in ways to teach given the varieties of technology enhanced learning that are becoming available. Hence it is important to identify a set of procedural principles that might be used to devise a technology curriculum and guide its implementation. This will be addressed in the next section.

## Three Procedural Principles to Inform the Development of a Technology Curriculum

For this purpose I have identified the following three key procedural principles:

- Procedural principle 1—being true to the nature of technology.

  This creates boundaries for the endeavor, avoiding confusion with other areas of the curriculum while at the same time enabling relationships with them. However, it is not simple as ideas concerning the nature of technology are contested (see Chap. 2, this volume, by Steve Keirl).
- Procedural principle 2—developing a perspective on technology.

  It is important the pupils develop a view of technology and how it might be used. Such views should be developed through discussion and reflection as opposed to being taught through instruction (see Chap. 3, this volume, by Marilyn Fleer).
- Procedural principle 3—enabling technological capability.

  It is important that pupils experience what it means to 'do' technology as opposed to just learning about technology. This implies that pupils will devise and produce technological outcomes in a variety of forms (see Chap. 5, this volume, by David Mioduser).

The three procedural principles identified here can enable technology educators to develop curricula in which they can be confident from at least four perspectives. First, it will be valid in that it deals with technology in ways that recognise the nature of technology. Second, it will enable young people to take a critical yet constructive view of technological activity as it unfolds in society around them. Third, it will enable young people to be technological themselves using a variety of approaches and tools, both analogue and digital. Fourth, it will have the capacity to incorporate new and emerging technologies as they are invented and developed in the world outside school. In short the procedural principle approach will give rise to technology curricula that can be justified as part of a technology education for ALL young people.

There are of course other procedural principles that could be invoked to inform the development of a technology curriculum, for example, it must be possible to assess the curriculum that is developed, the developed curriculum must be affordable within school budgets, the developed curriculum must be 'teachable' by the majority of teachers, etc. However these principles, while important, are not specific to technology education. They operate at a lower level of significance. Hence I have restricted the discussion to the three principles I think are most important.

Each procedural principle will be considered in turn although it is important to realise that in using them to develop a technology curriculum they will interact with one another in a dynamic way, each informing to a greater or lesser extent any curriculum consideration.

## Being True to the Nature of Technology

Technology is not easy to define, as different philosophical positions lead to different definitions. Kelly (2010) in his provocative book *What technology wants* discusses the idea of autonomous technology in terms of three interacting influences.

> The primary driver is pre-ordained development—what technology wants. The second driver is the influence of technological history, the gravity of the past, as in the way the size of a horse's yoke determines the size of a space rocket. The third force is society's collective free will in shaping the technium, or our choices. (p. 181)

From this perspective it appears that the influence that mitigates against technological inevitability is the smallest of these influences. He compounds this position by describing technological development in terms of a set of trends that contribute to the expression of particular technologies and how they might progress. In this set he includes increasing sentience, which may give cause for concern given that deeply embedded in popular culture is the idea of machines becoming self aware and either dominating human life as in the film Metropolis (1927) or deciding that humanity is antithetical to its own existence as in the Terminator films (1984, 1991, 2003 and 2009) and actively waging war on humanity.

Nye (2006) rejects this idea:

> From the vantage point of the present, it may seem that technologies are deterministic. But this view is incorrect no matter how plausible it may seem. Cultures select and shape technologies, not the other way around ...A more useful concept than determinism is technological momentum, which acknowledges that once a system such as a railroad or an electrical grid has been designed to certain specifications and put in place it has a rigidity and direction that can seem deterministic to those who use them. (Location 2075 of 2662)

Arthur (2009) takes a different starting point in considering the nature of technology and the way it evolves. He argues that technology can be seen as the exploitation of phenomena revealed by science. He rejects a simplistic 'technology is applied science' view but is adamant that it is from the discovery and understanding of phenomena that technologies spring.

He notes that

> It should be clear that technologies cannot exist without phenomena. But the reverse is not true. Phenomena purely in themselves have nothing to do with technology. They simply exist in our world (the physical ones at least) and we have no control over their form and existence. All we can do is use them where usable. Had our species been born into a universe with different phenomena we would have developed different technologies. And had we uncovered phenomena over historical times in a different sequence, we would have developed different technologies. (p. 66)

John Naughton (1994) adds further weight to the rejection of a simplistic applied science view of technology when he writes that technology always involves "ways of doing things. . . a complex interaction between people and social structures on the one hand and machines on the other" (p. 12). John's description immediately, and to my mind rightly, complicates the technology curriculum in that a consideration of machines, which many would see as a basis for a technology curriculum, becomes insufficient.

The work of Wynn Harlen and colleagues (e.g., Harlen 2010) in developing statements of content for science education that were true to the nature of the subject may provide us with a useful model. They divided the content into ideas **about** science and ideas **of** science. What might be developed if we adopted such an approach for technology education? Here are my suggestions:

Ideas **about** technology might include

- Through technology people develop technologies and products to intervene in the natural and made worlds.
- Technology uses knowledge, skills and understanding from a wide range of sources, especially but not exclusively science and mathematics.
- There are always many possible and valid solutions to technological and product development challenges, some of which will meet these challenges better than others.
- The worth of technologies and products developed by people is a matter of judgement.
- Technologies and products always have unintended consequences beyond intended benefit which cannot be fully predicted by those who develop them.

Ideas **of** technology might include

- Knowledge of materials

Technological activity requires the use of materials. And if someone is going to use materials he or she will need to know something about them. So what needs to be known? Clearly the idea of properties, with different materials having different properties, is essential. Given the importance of eco footprints, it will be useful to know something about sources of materials and how they are refined to the state where they are useful. And given the finite nature of the material world it will be useful to know something about the estimated reserves of materials, especially those that are particularly useful and in short supply. This can be listed as:

Sources
Properties
Footprint
Longevity

- Knowledge of manufacturing

The next step of course is to be able to do something with these materials, so manufacturing is an important idea of technology. In broad-sweep terms manufacturing can be divided into four main methods—subtraction, addition, forming and assembly—and overlaid on each of these are methods of finishing. At the moment addition is receiving considerable attention as additive manufacture is being used to produce items of both simplicity and complexity at very different scales to the point where it will almost certainly be possible to 'print' organs for transplant. This important idea of technology can be subdivided as:

By subtraction
By addition
By forming
By assembly
With finishing

- Knowledge of functionality

Most of the made world has to 'work' so some knowledge of achieving functionality is required. Three categories spring to mind: powering, controlling and structuring. Controlling is moving on in leaps and bounds with the embedding of electronic intelligence into everyday products becoming commonplace and the technology to achieve this is within the reach of schools through microcontrollers such as picaxe and arduino. So this important idea of technology can be sub divided as:

Powering
Controlling
Structuring

- Knowledge of design

Very little of the made world comes into existence except through designing. So knowledge of design is crucial, but teaching designing has long been seen as the Achilles heel of the subject. Four broad methods will be needed: identifying peoples' needs and wants, identifying market opportunities, generating and developing design ideas, and evaluating design ideas. This set of methods taken together and used sensibly will enable people to envisage outcomes that do not as yet exist, and create these outcomes through choosing and using materials and embedding function. So this important idea of technology can be sub divided as:

Identifying peoples' needs and wants
Identifying market opportunities
Generating and developing design ideas
Evaluating design ideas

- Knowledge of critique with regard to impact

The question that immediately follows is to what extent are these outcomes of worth? How do they affect the lives of those who use them and those that make them? How do they affect the planet? Here we immediately see the need for critique. This is different from evaluation as defined in 'evaluating design ideas'. Two broad areas of critique are stewardship and justice. Critiquing for stewardship involves considering life cycle analysis and speculating about different economic models—the currently predominant linear economy and the circular economy as espoused by, for example, the Ellen MacArthur Foundation (2012a, b). In a just world all people should be able to live in freedom from hunger and fear and have shelter from harm. They should have opportunities to pursue happiness and make the best of their lives. The made world full of deliberately designed products, environments and systems must be held to account by critique. So critiquing the outcomes of others is an important pupil activity. This important idea of technology can be sub divided as:

For justice
For stewardship

Within such a framework, 'ideas about technology' would mainly inform the development of a perspective on technology while the 'ideas of technology' would be essential for enabling technological capability. It must be acknowledged that these 'ideas of technology' will lead to a view of capability that is limited to the extent that it will manifest itself mainly through pupils' designing and making activities. However, despite this limitation, these complementary aspects of technology can be used to shape a technology curriculum such that it is true to the nature of technology.

## *Developing a Perspective on Technology*

The cultural and democratic arguments for teaching technology cited earlier in this chapter have as one of their aims giving young people insight into 'how technology works' such that they develop a constructively critical view of technology, do not become alienated from the technologically-based society in which they live, and are able to consider how technology might be used to provide products and systems that help create the sort of society in which they wish to live. Underpinning this is the idea of giving young people a perspective on technology which meets these aims.

An important concept for understanding the way technology works is that of disruptive innovation as defined by Christensen (2012): novel technologies that disrupt prevailing markets. The way in which transport was mechanised by the application of the internal combustion engine to vehicles that had hitherto been horse drawn was a revolutionary technological innovation. But it was not, at its inception, a disruptive innovation because early automobiles were expensive luxury items that did not disrupt the market for horse drawn vehicles. The market for transportation essentially remained intact until the debut of the lower priced Ford Model T in 1908. Henry Ford (1922) indicated his intention to be disruptive as follows:

> I will build a car for the great multitude. It will be large enough for the family, but small enough for the individual to run and care for. It will be constructed of the best materials, by the best men to be hired, after the simplest designs that modern engineering can devise. But it will be so low in price that no man making a good salary will be unable to own one—and enjoy with his family the blessing of hours of pleasure in God's great open spaces. (p. 73)

The means to mass-produce automobiles was the disruptive innovation because it changed the transportation market. The automobile, by itself, was not. It was the development of this production system that enabled the profitable manufacture of large numbers of vehicles that were affordable both in terms of purchase price and running costs, giving rise to a transport system that completely revolutionised our way of life. To accommodate the needs of the motorist (and to provide for movement of goods by lorries and tankers), a large network of roads and motorways has developed. The use of motor vehicles on this transport network contributes significantly to pollution of the atmosphere and global warming. Learning to drive and acquiring a motorcar have become a rite of passage for most young adults, male and female, in many countries. The opportunity to move from your place of birth to new and different places, to gain employment, to meet new people, to form friendships and relationships is facilitated by the motorcar. However, this physical and social mobility can have a deleterious effect on small, localised communities. Since the first road-crash fatality in 1896, motor vehicles have claimed an estimated 30 million lives globally. On average, someone dies in a motor-vehicle crash each minute in the world.

When the motorcar was invented and the automobile industry was born, no one envisaged that subsequent design iterations would be responsible for environmental damage, social upheaval, and a colossal death toll. What would Henry Ford have said if it had been suggested to him that his manufacturing innovation would harm the planet, erode family values, and kill millions of people? He would probably have been incredulous. He might have argued that society would step in and stop all those dreadful consequences by managing the way this new technology would be used. But he would have been wrong. Of course, Henry Ford and his designers and engineers were not of malign intent. They saw what they were doing as providing considerable benefit to many people. And, in that respect, they were correct. The motorcar has been highly beneficial to many individuals, communities, and societies—but at a cost: the cost of impact beyond intended benefit.

It is instructive to go back even further in time and consider the impact of a very simple technological development—the needle. The first evidence of the use of needles is circa 40,000 years ago. The ability to sew animal skins together to produce clothing and tent-like structures had a major impact on the lives of early humans. The use of the needle and thread made one's life not only more comfortable but increased one's chance of survival. Couple this with the invention of felting and weaving for the production of cloth and the humble needle can be seen as a revolutionary technological development. It was not disruptive in the sense identified by Christensen (2012) in that it did not disrupt the markets for clothing and tent production, as these markets did not yet exist. In fact one might argue that the needle created these markets. It was not until the late eighteenth century that the needle became mechanised with the invention of the sewing machine. Initially these were hand driven, later by foot pedal and later still by the addition of an electric motor. But a significant limiting factor in the use of the sewing machine, whether hand, foot or electric motor driven, is that it requires a human operator. Although computer-aided embroidery is now well established, and while the embroidery itself is computer-driven, the placing of the fabric to be embroidered on the machine is still carried out by humans. At the time of writing the robots so ubiquitous now in car manufacturing cannot handle fabric. This requires the dexterity of humans. It remains to be seen whether fabric-handling robots will be developed, but even if they are they will not be disruptive in that they cause the clothing market to be replaced or significantly reduced by another market. However it is noteworthy that the use of robots in a wide range of manufacturing is increasing and that this is seen as reducing the need for human labour (Finkelstein 2013)—an important consideration when developing a perspective on technology.

In developing a perspective on technology it will be important to engage young people with technological trajectories into the future as opposed to simply providing information of past trajectories up until the present, although appreciation of technology history is not seen as without merit. Witness the inclusion in recent design & technology National Curricula in England of statements such as "... they evaluate present and past design and technology, and its uses and effects" (QCA 2007) and "Through the evaluation of past and present design and technology, they develop a critical understanding of its impact on daily life and the wider world" (Department for Education 2013a).

Might it be possible for pupils at school to speculate in a rational way about the future uses of technology and will they respond positively and enthusiastically? There is some encouraging evidence from the Young Foresight project (Barlex 2012) that this might be the case. The Young Foresight project required pupils to work collaboratively in designing but not making products and services. The project identified four factors that teachers should encourage their pupils to take into account:

- The technology that is available for use. This should be a new and/or emerging technology and be concerned primarily with how the new product or service will work. Pupils should not concern themselves with manufacture.

- The society in which the technology will be used. This will be concerned with the prevailing values of the society, what is thought to be important and worthwhile. This will govern whether a particular application of technology will be welcomed and supported.
- The needs and wants of the people who might use the product or service. If the product does not meet the needs and wants of a sufficiently large number of people then it will not be successful.
- The market that might exist or could be created for the products or services. Ideally, the market should be one with the potential to grow, one that will last, and one that adapts to engage with developments in technology and changes in society.

Clearly, these factors interact with one another and influence the sorts of products and services that can be developed and will be successful. Considering these parameters, unencumbered by the necessity of making the proposed designs, enables pupils to be creative and develop highly original conceptual design-proposals. Some of the products and services devised by groups of Year 9 pupils in response to the challenge of utilising the stress sensitive conductor QTC (Quantum Tunnelling Composite) included the following:

- clothing that changes colour as you dance
- car tyres that sense their internal pressure
- an epileptic fit detector
- a self-weighing suitcase
- an arthritis treatment device
- keep fit apparatus
- a depth sensitive submersible
- an internal heart beat monitor.

The development and justification for products and services derived from new and emerging technologies could be a first step in developing a perspective on technology and the Young Foresight approach provides one way of doing this. To enhance this perspective it would be necessary for pupils to put their suggestions into an historical context: what, if any, similar products or services had existed before, how were they received by society, and what impacts did they have on society? It would also be important for pupils to speculate about impacts beyond the intended benefits that might arise if their ideas had wide implementation. Three important features of this approach to developing a perspective on technology are (a) it enables the curriculum to keep pace with the technological developments and innovations taking place in the world outside school, (b) pupils can choose for themselves which new and emerging technologies they wish to consider, and (c) pupils can be encouraged to actively speculate about the way new technologies might play out in alternative technological futures.

## *Enabling Technological Capability*

This principle is important because it captures an essential feature of technological activity—that its main concern is with intervention as opposed to understanding; a key difference between science and technology. Drexler (2013) captures this well when he describes the difference between engineering design and scientific enquiry. Whereas science begins with a physical system (the object of study) and moves towards an abstract model (theory), engineering starts with an abstract model (a design concept) and moves towards a physical system (useful product). David Layton was particularly insightful here when he wrote that the primary purpose of design & technology education is to develop in pupils the ability to "intervene effectively and creatively in the made world" (Department for Education and Science and the Welsh Office 1988). Kimbell and Perry (2001) captured this requirement as follows:

> The real products of design & technology are empowered youngsters; capable of taking projects from inception to delivery; creatively intervening to improve the made world; entrepreneurially managing their resources; capably integrating knowledge across multiple domains; sensitively optimising the values of those concerned; and confidently working alone and in teams. (p. 19)

The comment "sensitively optimising the values of those concerned" is particularly important since the technical content identified in 'ideas of technology'—knowledge of materials, manufacturing and functionality—does not exist in a vacuum. When operationalised through designing, value judgements have to be made as to what is worth doing with technical know how and what the consequences of doing such things might be. This is of course reflected strongly in the last two statements in ideas 'about' technology and in the inclusion of knowledge of critique in 'ideas of technology' (see the previous section).

Most technology curricula enable technological capability to some extent through activities in which pupils design what they are going to make and then make what they have designed. This is often seen as the heartland of technology education, although it does not reflect the reality of technological activity in the world outside school, where those who design artefacts are usually not those who manufacture them. As has already been acknowledged, this approach can be seen as a limited view of capability, but it terms of feasible pupil activity it does embody 'intervention' in a powerful way.

Underpinning the activity is the act of designing, which featured significantly in 'ideas of technology'. Although making the designs of others is not without educational worth (see Barlex 2011a), the essence of capability comes from pupils conceiving their own designs that they then realise. Hence it is important to pay considerable attention to the role of designing in enabling technological capability.

The decision making that students have to undertake when they are designing and making has been described as involving five key areas of interdependent design decision:

- conceptual (overall purpose of the design, the sort of product that it will be)
- technical (how the design will work)
- aesthetic (what the design will look like)
- constructional (how the design will be put together)
- marketing (who the design is for, where it will be used, how it will be sold).

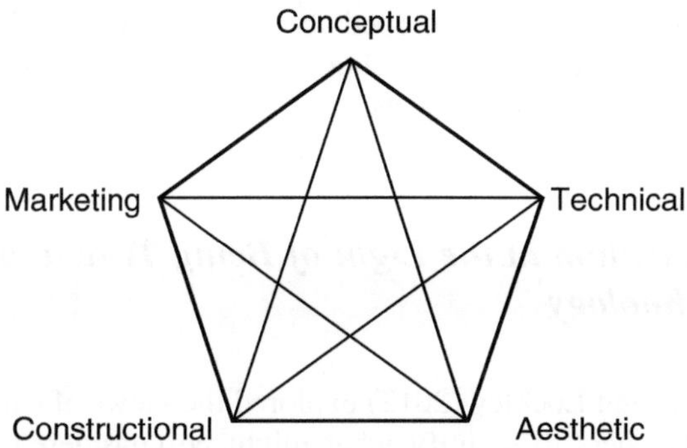


**Fig. 8.1** The design decision pentagon

This approach can be represented visually as a pentagon diagram shown in Fig. 8.1. The interdependence of the areas is an important feature of design decisions; hence the lines connect each vertex of the pentagon to all the other vertices. A change of decision within one area will affect some, if not all, design decisions that are made within the others. Although the teacher usually identifies the sort of product the students will be designing and making, which makes it very difficult for students to engage in conceptual design, there are still many opportunities for making design decisions in the other areas. It is the juggling of these various decisions to arrive at a coherent design proposal that can then be realised to the point of a fully working prototype that provides the act of designing and making with such intellectual rigour and educational worth that it is seen as an essential aspect of technology education.

The pentagon diagram can be further developed by adding three important supplementary questions when considering the combined effect of the design decisions that result in a particular product: What is the social impact of the product? What is the economic impact of the product? What is the environmental impact of the product? (See Barlex 2011b.)

It is worth noting that enabling technological capability can be situated in contexts in which there are inequalities to be resolved. In short, issues of justice are an important feature of critique noted in 'ideas of technology'. The writing of Emily Pilloton (2009) in *Design Revolution* is inspirational in this regard. Emily adopts an unashamedly social activist position with regard to design, and to my mind by implication technology.

It is through design and subsequent making that technologies of various sorts are developed and then utilised, be it for good or ill. Engaging students in using technology for societal improvement through designing and making is challenging, but can be defended by applying the procedural principle 'enabling technological capability' to curriculum development and implementation.

## Possible Futures for Technology Education

In this section I consider possible future technology curricula in the light of each of the three procedural principles developed above. I also consider briefly the overall future of technology education in the school curriculum, particularly in terms of its relationships with other school subjects.

### *A Future Curriculum in the Light of Being True to the Nature of Technology*

Recently, Williams and Lockley (2012) explored the views of early career science and technology teachers to identify what might be considered 'enduring ideas' within the subjects they taught. Interestingly, this research revealed that while the science teachers had little difficulty in identifying such ideas this was not the case for the technology teachers. The authors noted that this may be in part due to the extensive place of procedural knowledge in technology but also that technology has no commonly agreed upon epistemology. A similar enquiry that I carried out with a colleague (Barlex and Steeg 2013) with trainee design & technology teachers at two universities in England revealed similar findings. The responses of the trainees indicated that there was little uniformity concerning enduring ideas and that there was considerable difference between their individual perceptions of the nature of the subject.

To those of us who know design & technology teachers this might not come as a surprise, but the lack of unanimity is a cause of great concern. If the subject community has difficulty in reaching agreement on subject content then it is small wonder that those outside the subject, especially government ministers and their officials, find the subject difficult to fathom. Hence the procedural principle of being true to the nature of technology is of fundamental importance to the future of technology curricula.

It is noteworthy that understanding the nature of technology is an explicit feature of the New Zealand Technology Curriculum (Ministry of Education 2007), while such understanding is only implicit in the curriculum of England (Department for Education 2013a). At a research level, attempts are being made to establish a more coherent view across various communities of practice. These include the Delphi study carried out by de Vries et al. (2009) and the recent publication of *New principles for Design & Technology in the National Curriculum* by Education for Engineering (E4E 2013). Clearly technology curricula in different countries will need to respond to their particular cultural contexts (see Chap. 4, this volume) and it would be inappropriate for orthodoxy to become uniformity, but some generally accepted view is required if the subject is to be understood and valued.

A small case study that I recently conducted indicates that those training to be design & technology teachers respond positively to the complex and contested nature of technology (Barlex 2011b). The students were challenged with the following questions:

- Is technology autonomous and beyond our control or is technology under human control?
- Does technology control us or do we control technology?
- Is technology value-neutral or does it have implicit values?
- Does the availability of technology change human behaviour?
- Who decides which technologies are developed?
- Who decides which technologies are adopted?

These prompts led to wide-ranging discussions and an overall agreement that the nature of technology should feature in school design & technology curricula. While there were differences of opinion as to when such consideration should be introduced, the overwhelming view was that to avoid such a consideration would be to the detriment of the subject and compound the erroneous view of the Expert Panel that the subject lacked disciplinary coherence.

## *A Future Curriculum in the Light of Developing a Perspective on Technology*

Helping young people to develop a perspective on technology requires them to appreciate values that empower them to critique, a facility that Layton (1995) has argued is an essential feature of technology education. The specific inclusion of critique in 'ideas of technology' and the place of both intended and unintended impacts in 'ideas about technology' highlight critique as an increasingly important aspect in technology education.

It is only relatively recently that concern has been raised about the impact of technological activity on the environment, although some prescient voices such as Rachel Carson (1965) in *Silent Spring* began making the case much earlier. The recent pronouncements of the Intergovernmental Panel on Climate Change (IPCC) indicate that the science underpinning concerns about global warming is robust and that such warming will have considerable effects on the weather patterns of the planet:

> Human influence on the climate system is clear. This is evident from the increasing greenhouse gas concentrations in the atmosphere, positive radiative forcing, observed warming, and understanding of the climate system. (IPCC 2013, p. 13)

It is also generally accepted that the rate at which we are consuming finite resources is unsustainable. Both global warming and the rate of finite resource consumption can be laid at the door of technological activity, although it is reasonable to describe these as unintended consequences or impact beyond intended benefit. Unintended or not, however, they are the cause of considerable concern and now most school technology curricula give significant place to what might be termed 'education for sustainable development'.

Pavlova and Pitt (2007) believe that it is vital to avoid the implicit suggestion that sustainability is just another thing to think about in technology education, but that it should be a major consideration throughout the entirety of a school technology curriculum. However they are wary of preaching to achieve any significant change:

> There is a danger, however, that students (many of whom are products of a materialistic, individualistic, hedonistic, high-consumption-and-bugger-the- consequences, instant gratification society) will be bored out of their heads by overt moralising. (p. 86)

This poses a significant challenge to a school technology curriculum, as individual students and their families have little power over which technologies are created and, if they are to take part in society, little power over which technologies they use. To achieve a society in which a prevailing value is 'stewardship of planet earth' thinkers such as Braungart and McDonough (2009) and Webster and Johnson (2008) are arguing for a radical shift in the way technological activity is carried out. They argue that we should move from the current linear economy to a circular economy, which mirrors the way nature operates. In a linear economy finite materials are gathered from the natural world, turned into goods, which are used and then disposed of. In most cases this disposal leads to the materials becoming part of landfill. Hence, society will ultimately run out of these useful materials. The use of recycling, which is not an efficient process, only delays the time until the materials run out. It is not a long-term solution. In a circular economy, by contrast, the concept of waste is eliminated. Materials that would be seen as waste in a linear economy become the feedstock for other processes, mirroring the way materials in the natural world are always part of cycles that move them through a variety of linked ecosystems.

To become effective, products and production systems will need to be redesigned to meet this circular economy approach. This approach has been adopted by the Ellen MacArthur Foundation, which is actively and successfully lobbying both politicians and business leaders about the commercial and environmental benefits of moving to a circular economy. The Foundation has produced two significant reports (2012, 2013) identifying the considerable business opportunities in moving towards a circular economy. By responding to the procedural principle 'developing a perspective on technology', the curriculum will be able to consider the way technological activity is carried out in the world outside school and how different economic models can give rise to different futures.

A second dimension of critique that was identified in 'ideas of technology' was justice. It is clear that new and emerging technologies will have considerable impact on society in ways that affect the lives of individuals and communities. Which technologies are developed, the ways in which they might be deployed, and to whom they are made available are all important considerations for developing a perspective on technology. I think it is likely that technology curricula will respond to the 'perspective' procedural principle not only in terms of stewardship but also justice. This is starting in a small way in England through a curriculum development project concerned with disruptive technologies (Barlex et al. 2013) which will consider how the following technologies might be introduced into the school technology curriculum through a collaboration between school teachers, teacher trainers and those engaged in related research and development:

- Additive manufacturing
- Artificial intelligence
- Augmented reality
- Big data
- Intelligent matter
- Internet of things
- Neurotechnology
- Robotics
- Synthetic biology

Reasons for including aspects of disruptive technologies in the design & technology curriculum include:

- To help learners in design & technology engage in futures thinking.
- To engage learners with the ways in which technology leads to change.
- To help teachers understand what critiquing disruptive technologies offers to design & technology education.
- To begin to equip learners for a world in which many technologies are rapidly becoming more democratised and more available to the masses.
- To start to unpick, with learners, how new affordances will redistribute social, moral, environmental, financial, etc. responsibilities.

Here again we see how the adoption of the procedural principle approach outlined above facilitates the introduction of new elements into the technology curriculum.

## *A Future Curriculum in the Light of Enabling Technological Capability*

The tools available to young people engaged in technological activity at school are becoming more sophisticated. The E4E (2013) publication *New principles for Design & Technology in the National Curriculum* deliberately formulated its approach in terms of a toolbox containing features that pupils could use in developing and demonstrating their technological capability. The availability of a wide range of CAD software with significant tutorial support, some freely available as in the case of SketchUp and Autodesk, enables young people to develop design proposals that range from the simple to the sophisticated.

An issue for those teaching pupils to use such software as part of the technology curriculum is of course the worth of what is being designed. My view here is that it is important to generate ideas of worth. A question that should be addressed by both teachers and pupils is, "To what ends are we deploying our technological capability?" This might be rephrased as, "Is what we are designing worth making?"

An additional issue in pursuing the procedural principle of enabling technological capability is the extent to which pupils have a voice in deciding the nature of the tasks they tackle. Brundrett and Silcock (2002) raise this issue when they discuss the idea of a partnership curriculum in which teachers and pupils are engaged in a process of negotiation. Recently, the Design & Technology Association in England posted on their website a set of open starting points for designing and making electronic products (Design and Technology Association 2010). These starting points are available as visual brainstorms that the teacher can use with the class to explore the context and identify many different sorts of electronic product that could be designed and made in response. These open starting points provide the opportunity to give pupils a voice as to what sort of product they want to design and make. The exact nature of the products designed and made will depend on the age and previous experience of the pupils and the resources available in the school, but giving the pupils a voice will increase their influence on the curriculum and provide greater ownership, which is likely to increase their motivation. It is easy to see how such negotiation could be enhanced through using social media, and I suspect that tasks of worth identified through teacher-pupil negotiation will feature more prominently in future technology curricula.

Computer-controlled tools for manufacture are now increasingly available in schools in England. The laser cutter is almost ubiquitous, with some schools having two or three machines. Since their introduction they have become cheaper, more reliable and durable. They allow 'cutting out' of complex shapes from a range of sheet materials, including textiles, that could never be achieved by pupils using conventional hand or machine tools. The 3D printer is also becoming more widely available and while it has yet to achieve the reliability and durability of the laser cutter it is coming down in price. There is also the possibility that some pupils will have access to 3D printers at home—there is a burgeoning hobby market—or via local stores which provide 3D printing facilities. So it is likely that in the near future that pupils will have access to these modern manufacturing tools that rely on digital design information. This will place the designing and making of pupils in the context of digital rights management issues as applied to products, an arena of activity that some are arguing should be tackled by Apple given their success in handling the issue with regard to music (Cilderman 2013).

It is not only in the production of form that pupils have access to sophisticated tools. The availability of inexpensive microcontrollers such as PICAXE and Arduino along with significant free-to-download tutorial support puts the designing of products with embedded intelligence easily within the grasp of pupils aged 11–14 years. The development of even more accessible programming languages based on, for example, Scratch as being developed by MIT, will enable pupil technological capability to take on a contemporary dimension unimaginable in the days of hard-wired electronics.

Here it can be seen see that pursuing the procedural principle of enabling technological capability with regard to designing and making products of worth makes it possible to give pupils access to sophisticated computer-based design and manufacture tools in parallel with the availability of similar tools to technology professionals. This provides an opportunity to continually update the technology curriculum to keep it in line with parallel developments in the world outside school. An important caveat here is that the newly available design tools should not be used for their own sake but in pursuit of tasks of worth, which can increasingly be identified through teacher-pupil negotiation.

## *Identity and Relationships*

The sense of technology's own identity is important for two reasons. First, it becomes possible to establish technology as a subject in its own right within the school curriculum. This is necessary for the subject to be recognised, understood and valued by those responsible for the rest of the curriculum. This leads to the second reason: the nature of technology is such that it calls upon knowledge from other curriculum areas. Relationships between technology and other subjects that are mutually beneficial can only be developed and sustained if technology has a clear sense of self and its own unique contribution to education. Without this, relationships with other subjects become ones in which their educational goals dominate and technology is always the lesser partner, sometimes to the extent that it is no longer true to its unique nature. This is particularly in the case of relationships with science when technology becomes reduced to applied science. This need not be the case, and Banks and Barlex (2013) have identified a wide range of activities and approaches which exemplify the synergy and enhanced learning that can be achieved through so-called STEM activities without compromising the integrity of the contributing subjects. Chap. 10 of this volume also explores possibilities for alignment between technology and other subjects, and the potential implications.

Here it must be acknowledged that in the short term technology is unlikely to become a 'gatekeeper' subject like science or mathematics although recently the government in England has indicated that a qualification in design & technology can contribute to the school performance measures used as an accountability system for secondary schools (Department for Education 2013b). In New Zealand, a significant political step was the inclusion of technology education into the canon of subjects recognised for University entrance. However, in many countries it is unlikely that achieving qualifications in school technology courses will become an essential pre-requisite for entry into particular forms of further or higher education or occupation. This does not in any sense deny its value as an essential component in the general education for all pupils, but it does mean that those who teach technology in these countries will have to be pro-active in promoting its benefits and ensuring it is valued by those who have influence in the curriculum.

Some argue that limiting links to mathematics and science alone, as in many STEM proposals, is insufficient. Hence there is a small but growing STEAM movement— where STEAM stands for science, technology, engineering, arts and mathematics. White (2012) provides a rationale for this widening of links to the arts as follows:

- Arts education is a key to creativity.
- Creativity is an essential component of, and spurs innovation.
- Innovation is agreed to be necessary to create new industries in the future.
- New industries, with their jobs, are the basis of our future economic wellbeing.
- A win-win situation—low cost, job growth and insuring the future.

He cites US Education Secretary Arne Duncan speaking to the Arts Education Partnership National Forum in April 2010:

> The arts can no longer be treated as a frill . . . arts education is essential to stimulating the creativity and innovation that will prove critical to young Americans competing in a global economy. . ..

I would question this highly utilitarian approach to arts education but I think that another rationale can be developed based on our second procedural principle of developing a perspective on technology. This is exemplified by the work of James Bridle (2013), an artist and an activist, who hopes his work calls into question our sometimes blind reliance on technological systems with which we come into contact every day. Hence he has engaged with the debate on the use of drones for covert warfare through his project *Dronestegram*, a social media tool that identifies recent areas of drone strikes, and *Drone Shadow*, a full-scale outline of a drone painted on the sidewalk outside of the Corcoran Gallery across the street from the White House. I think this provides an interesting model for developing critique, a key feature of developing a technological perspective.

For teachers who wish to develop their own technology curricula the three procedural principles identified provide signposts for their endeavours. Teachers can identify areas of learning and associated activities they think will be appropriate and use the principles to scrutinise their validity. In a similar way the three principles give technology educators the tools with which to consider and critique any technology curricula that might be espoused by governments or agencies seeking to modify current technology curricula or to introduce technology curricula for the first time. Once the worth of any proposed changes has been clarified and acknowledged the extent to which technology teachers are able to respond positively to any worthwhile changes can be discussed. This allows teachers to identify and lobby for relevant professional development. It also allows teacher educators to modify their offerings in initial teacher training.

In looking at future technology curricula it is worth considering the views of Keri Facer (2011), who challenges the prevailing concept of the school. In her book, *Learning futures, education, technology and social change*, Facer reconceptualises schools as places where communities build their *own* future. Keri is particularly concerned that advances in new and emerging technologies will have significant and as yet unknown impacts on our society and by implication on education. At the end of the book she outlines nine conditions to enable future-building schools. Of these, one is particularly relevant to technology education: develop an ethical code for the educational use of digital and biotechnologies. Here Facer argues that schools should no longer be recipients of new socio-technical practices developed in the world outside school but could rewrite the relationship between education and socio-technical change as one of active design, critique and engagement. Facer envisages these changes as having the potential to influence the nature of schools within 20 years. Just imagine what that would mean for teaching and learning technology.

## Summary

This chapter began by asking the question "How does a curriculum come into being?" Through exploration and discussion, the first part of the chapter moved to the position where it was necessary to identify procedural principles to inform the development of a technology curriculum. The second part of the chapter introduced the following as three procedural principles: being true to the nature of technology, developing a perspective on technology, and enabling technological capability. These were then each discussed in some detail. Finally, the discussion of these principles was used to explore possible futures for technology education and to consider the relationships between technology and other school subjects.

A key question must be, "Who gains and who loses from the use of the three procedural principles identified and argued for in this chapter?" I believe that the use of the procedural principle approach to technology curriculum development will empower teachers to devise, justify and implement programmes of study that are robust and can withstand scrutiny from those who might question the worth of technology education. Governments wishing to influence or dictate technology curricula will find it more difficult to go against the justified arguments of educational professionals. While this might seem to put such governments in a 'losing position' the opposite is the case. Their thinking could become more informed through discourse with an articulate and knowledgeable profession. Those who will lose are those who adopt a narrow economic argument for technology education to the exclusion of wider considerations.

## References

Alexander, R. (1984). *Primary teaching*. London/New York: Holt, Rinehart & Winston.

Arthur, W. B. (2009). *The nature of technology*. London: Allen Lane.

Banks, F., & Barlex, D. (2013). *Teaching STEM in secondary schools. Helping teachers meet the challenge*. London: Routledge.

Barlex, D. (2007). Assessing capability in design & technology: The case for a minimally invasive approach. *Design and Technology Education: An International Journal, 12*(2), 9–56.

Barlex, D. (2011a). Dear Minister, This is why design & technology is a very important subject in the school curriculum. *Design and Technology Education: An International Journal, 16*(3), 9–18.

Barlex, D. (2011b). Teaching young people about the nature of technology. In K. Stables, C. Benson, & M. de Vries (Eds.), *PATT 25: CRIPT 8 Perspectives on learning in design & technology education* (pp. 66–75). London: TERU, Goldsmiths.

Barlex, D. (2012). The Young Foresight Project. A UK initiative in design creativity involving mentors from industry. In B. France & V. Compton (Eds.), *Bringing communities together: Connecting learners with scientists or technologists* (pp. 113–126). Rotterdam: Sense.

Barlex, D., & Steeg, T. (2013, March 7–9). *Student teachers' perceptions of 'enduring ideas' in design & technology*. Paper presented at PATT ITEEA Conference, Improving Technology and Engineering Education for all students: Research based plans for action. Columbus, OH, USA.

Barlex, D., Givens, N., & Steeg, T. (2013) Disruptive technologies: engaging teachers and secondary school students in emerging affordances. In P. J. Williams, & D. Gedera (Eds.), *PATT 27 Technology Education for the future: A play on sustainability* (pp. 35–43). New Zealand: University of Waikato; University of Canterbury.

Black, P. (2013). Personal communication, 16 July 2013.

Braungart, M., & McDonough, W. (2009). *Cradle to cradle*. London: Vintage.

Bridle, J. (2013). *Drone art: James Bridle poses questions in White House's shadow*. Accessed 28 June 2013 from: http://www.bbc.co.uk/news/magazine-22997396

Brundrett, M., & Silcock, P. (2002). *Achieving competence, success and excellence in teaching*. London: Routledge/Falmer.

Bruner, J., & Haset, H. (Eds.). (1987). *Making sense. The child's construction of the world*. London: Methuen.

Carson, R. (1965). *Silent spring*. London: Penguin.

Christensen, C. M. (2012). Disruptive innovation. In M. Soegaard, D. Soegaard, & F. Rikke (Eds.), *Encyclopedia of human-computer interaction*. Aarhus, Denmark: The Interaction-Design.org Foundation. Accessed 23 Nov 2013 from: http://www.interaction-design.org/encyclopedia/disruptive_innovation.html

Cilderman, M. (2013). *3D printing needs Apple and Apple needs 3D printing*. Accessed 25 June 2013 from: http://seekingalpha.com/article/1499652-3d-printing-needs-apple-and-apple-needs-3d-printing

De Vries, M., Hacker, M., & Rossouw, A. (2009). *Concepts and contexts in engineering and technology*. Delft: Delft University of Technology and Hofstra University.

Department for Education. (2012). *Expert panel terms of reference*. Accessed 25 June 2013 from: http://www.education.gov.uk/schools/teachingandlearning/curriculum/nationalcurriculum/a0073091/expert-panel-terms-of-reference

Department for Education. (2013a). *Design & Technology Programme of Study: KS3*. Accessed 21 Nov 2013 from: https://www.gov.uk/government/uploads/system/uploads/attachment_data/file/239089/SECONDARY_national_curriculum_-_Design_and_technology.pdf

Department for Education. (2013b). *Reforming the accountability system for secondary schools*. October 2013. England: Department for Education.

Department for Education and Science and Welsh Office. (1988). *National Curriculum Design and Technology working group interim report*. London: HMSO.

Design & Technology Association. (2010). I*nformation about open starting points*. Accessed 25 June 2013 from: http://www.ectcurriculum.org/index.php/starting-points

Doll, W. E. (1993). *A post-modern perspective on curriculum*. New York: Teachers College.

Drexler, K. E. (2013). *Radical abundance*. New York: Public Affairs.

Education for Engineering (E4E). (2013). *New principles for Design & Technology in the National Curriculum*. London: Royal Academy of Engineering.

Ellen MacArthur Foundation. (2012a). *Towards the circular economy Volume 1: Economic and business rationale for an accelerated transition*. Accessed 25 June 2013 from: http://www.ellenmacarthurfoundation.org/business/reports/ce2013

Ellen MacArthur Foundation. (2012b). *Towards the Circular Economy Volume 2: opportunities for the consumer goods sector*. Accessed 25 June 2013 from: http://www.ellenmacarthurfoundation.org/business/reports/ce2013

Esland, G. M. (1971). Teaching and learning as the organization of knowledge. In M. F. D. Young (Ed.), *Knowledge and control* (pp. 70–115). London: Collier Macmillan.
Facer, K. (2011). *Learning futures: Education, technology and social change*. London/New York: Routledge/Taylor and Francis Group.
Finkelstein, D. (2013). *Machines are becoming cheaper than labour*. Accessed 22 Nov 2013 from: http://www.thetimes.co.uk/tto/opinion/columnists/danielfinkelstein/article3914166.ece
Ford, H., with Crowther, S. (1922). *My life and work*. Garden City: Doubleday.
Harlen, W. (2010). *Principles and big ideas of science education*. Hatfield: Association for Science Education.
IPCC. (2013). *Climate change 2013. The physical science basis summary for policy makers*. Switzerland: IPCC.
Kelly, A. V. (2009). *The curriculum: Theory and practice*. London: Sage.
Kelly, K. (2010). *What technology wants*. New York: Viking.
Kimbell, R., & Perry, D. (2001). *Design and Technology in a knowledge economy*. London: Engineering Council.
Layton, D. (1995). Constructing and reconstructing school technology in England and Wales. *International Journal of Technology and Design Education, 5*(2), 89–118.
Ministry of Education. (2007). *The New Zealand Curriculum*. Wellington: Learning Media.
Naughton, J. (1994). What is 'technology'? In F. Banks (Ed.), *Teaching technology* (pp. 7–12). London: Routledge.
Nuffield Council on Bioethics. (2013). *Emerging biotechnologies: Technology, choice and the public good*. London: Nuffield Council on Bioethics.
Nye, D. E. (2006). *Technology matters*. Cambridge, MA: MIT.
O'Hear, A. (1992). The Victor Cook memorial lectures. In J. Haldane (Ed.), *Education, value and culture: The Victor Cook Memorial Lectures* (pp. 40–69). St. Andrews: University of St Andrews.
Pavlova, M., & Pitt, J. (2007). The place of sustainability in design & technology education. In D. Barlex (Ed.), *Design and technology for the next generation* (pp. 72–87). Shropshire: CliffeCo.
Peters, R. S. (1973). *The philosophy of education*. Oxford: Oxford University.
Pilloton, E. (2009). *Design revolution: 100 products that are changing people's lives*. London: Thanes and Hudson.
Qualifications and Curriculum Authority. (2007). *National Curriculum for design & technology*. Accessed 22 Nov 2013 from: www.qca.org.uk/curriculum
Stenhouse, L. (1975). *An introduction to curriculum research and development*. London: Heinemann.
Webster, K., & Johnson, C. (2008). *Sense & sustainability*. Accessed 25 June 2013 from: http://www.ellenmacarthurfoundation.org/education/resources/articles-and-books
White, H. (2012). Harvey White's website arguing for the involvement of the Arts in STEM can be found at: http://steam-notstem.com/
Williams, P. J., & Lockley, J. (2012). An analysis of PCK to elaborate the difference between scientific and technological knowledge. In T. Ginner, J. Helstrom, & M. Hulten (Eds.), *Technology education in the 21st century. Proceedings of the PATT 26 conference 2012* (pp. 468–477). Stockholm: Linkoping University.

# Chapter 9
# Developing a Deeper Understanding of Design in Technology Education

**David Spendlove**

The fundamental position for this chapter is that design is central to being human—everyone designs and engages in the process of designing. However, design is different in different contexts. While designing is an innate capacity, it is also a disciplined activity system located in industry, commerce, the arts and education, and has multiple definitions and uses. Common across contexts is that all design should be creative (though not all creativity involves design). Design also involves riskiness and uncertainty, and is an integral aspect of a sustainable economy, ethical lifestyle and the shaping of communities. While such views of design represent empowering learning opportunities for children associated areas such as creativity, riskiness and uncertainty have become increasingly marginalised in educational contexts demanding ever-greater accountability in terms of productivity and performativity. An opportunity does, however, arise when considering future-focussed technology education programmes that value 'design thinking' and how this can contribute to students' learning and 'being'.

## Introduction

In this chapter the aim is to offer a new perspective when considering the nature of 'design' as a phenomenon in 'Technology Education'. In doing so I am going to consider 'design' as a noun and as a verb. I will also consider the term 'designer' and processes that we call 'designing'. The reason for undertaking such a task is that these terms are far from stable—they are, like organic life forms, evolving in a rapidly changing world and as such their use in everyday vernacular, in different

---

D. Spendlove (✉)
Manchester Institute of Education, The University of Manchester, Manchester, UK
e-mail: David.Spendlove@Manchester.ac.uk

 
PJ. Williams et al. (eds.), *The Future of Technology Education*, Contemporary Issues in Technology Education, DOI 10.1007/978-981-287-170-1_9

cultural settings, by different groups and in different professional settings means that the word ‘design’ becomes adapted to its environment and often can be difficult to define.

To begin with, I want to state a fundamental position by asserting that design is central to being human. As such, we are all natural designers and we all engage in the process of designing. To clarify this position further it is important to note that when I propose we are natural designers I don’t suggest this in the context of a professional, vocational or commercial sense; rather, humans naturally identify problems and the seeking of resolutions is an innate function of evolution. We are naturally wired to seek challenges and to plan and ‘cognitively model’ imaginative solutions.

If design is therefore understood to be a naturally occurring human activity, we need to question why it should be taught and how should it be taught. In addition, while acknowledging that design is a universal attribute, it is important to recognise that when looking at design through the lens of professions—and then through a further lens of the Western orientated education systems of the world—the use of the term design requires a sharper focus if it is to be of value in helping us understand and engage with the phenomenon, particularly in an educational setting.

Inevitably, when making a statement that design is a central process of being human, there is a need to also consider what it means to be human—being human involves a series of values, beliefs and ethics that may not be the same for all of humanity. Equally, as design shapes the human world, we need to consider what this may mean in what is a rapidly changing world and how our wanting to change the world for the better (whatever this might actually mean) may distort the carefully balanced ecological systems that exist.

I want early on in this chapter to also offer a note of caution, as inevitably there is a degree of self-interest and self-preservation in this book. We are all interested in the future of technology education and have a vested interest in preserving it as we all passionately believe that it offers something special to the curriculum. However, we have to be careful that this self-interest does not become delusional and merely self-serving. The reason for making this point explicit is that if we are to embody the beliefs that we have about design then this means that we also have to consider and question the structure of the curriculum, the nature of education, the place of schools and learning establishments, teacher and pupil relationships, human fallibility, the nature of social class, politics, philosophy, religion and culture, as each area identified (and more) is encapsulated, either implicitly or explicitly, in every act of teaching and in every process of designing.

My point is that, like all good designers (whatever this may actually mean), educators have to be open minded and questioning in our search for solutions. To illustrate this I am reminded of the saying about ‘fish being the last to discover water’ and I think a way of understanding this chapter is about trying to help us all discover ‘the water’. In doing so there is a need to consider the development of design as an activity, not simply as a means for justifying self-survival (as in a place in the curriculum)—which is an extremely valid reason—but as a means to identify a set of qualities and dispositions which are an essential part of every child’s general education.

In summary, this chapter aims to explore what we mean by the term 'design' by challenging some of the assumptions that may exist, as well as considering some new ways of rationalising the existence of design within an educational context. By doing this, it is not an attempt to undermine or weaken the position of technology education and its relationship with design; quite the opposite. It is only by challenging and questioning assumptions about the location of design within technology education that a deeper understanding of its position in children's education can be developed and justified.

## Defining Design (or Not)

As previously stated, design is an innate human activity as well as a disciplined activity system located in industry, commerce, the arts and education, and everyday life. As such, it is not a stable term with an established definition, but the term has multiple uses depending upon the context in which it is used. For example, art and design, product design, systems design, engineering design, industrial design, and design and technology are all terms that use and define design in slightly different ways. As Love (2002) suggests, design has "different meanings in different domains, [is] used in different ways by researchers in the same domain, and [is] found in the literature referring to concepts at different levels of abstraction" (p. 347).

As such, unusually, design can also be used as a noun or a verb:

> The first usage is as a noun, connoting the field of design as a whole in a very general manner, as in the phrase: "Design is important to national economic competitiveness." The second usage is as a verb, meaning the action or thought involved in the act of designing. (Heskett 2001, p. 18)

In addition, Heskett identifies two further applications of the term, again as a noun but recognising design as "connoting a plan or intention" or meaning "the finished product". In other words, design can be used in any one of these four states, as well as in a series of combinations with other words to give very different meanings, yet there is a tendency to seamlessly move "from one meaning to another without distinction" (ibid).

In identifying the different uses of the term design it is useful to partially explore the evolution of the word, which might help explain why it is a difficult word to fully define. Originally, 'design' derived from the Latin *signum,* which means 'sign' and was used around the fourteenth century. Therefore, etymologically, *design* means to 'de-sign'. The evolution to 'design' encompassed a diverse range of activities and Flusser and Cullars (1995) make a useful contribution to the debate by suggesting that bourgeois culture made a distinction between the world of the arts and of technology, with design forming a bridge between the two: "Hence in contemporary life, *design* more or less indicates the site where art and technology (along with their respective evaluative and scientific ways of thinking) come together as equals, making a new form of culture possible" (p. 51). This definition of design as a noun sits within the field of 'design activities' rather than the process of designing domain.

Interestingly, Flusser and Cullars (1995) don't specifically identify aesthetics or products as a central feature of design, but determine design by its location (between art and technology) as a culture. This is worth reflecting upon further, as design—particularly in education and design-related professions—is often inextricably intertwined with manufactured products and aesthetics. However, the reconceiving of design as a broader, inclusive term where "design is increasingly understood in a much wider sense as the human capacity to plan and produce desired outcomes" (Mau 2007) and no longer associated simply with aesthetics, objects and manufacturing, poses new questions for developing design capability as an educational activity. Mau's definition is, however, consistent with an emerging shift in the professional design world to recognising design as an increasingly interdisciplinary activity that also involves immaterial artefacts such as systems, experiences, and environmental design.

This shift is important to note within the context of a broad general educational as there is a tendency for education to operate in a time lag, delivering an 'arts and craft movement' curriculum (which can offer significant benefits) while the 'real world' has moved on. This juxtaposition can, however, be rationalised by arguing that education is not merely vocational to serve the needs of industry—something which I would agree with. However, if design-related activities in an educational context are not delivered in a vocational way for vocational purposes then an alternative rationale has to be provided. Such a rationale can be contemplated when we consider the verb domain—doing design through action, thought and agency—and this is where the focus of the rest of this chapter will be.

## Creativity in Design (or Not)

The societal benefits of being creative within a technological experience provide a strong rationale for technology education as a significant force for children in enabling them to use their powers of creation to mould their environment, ultimately strengthening the stability of future societies. Thus, creativity is more than merely embellishment of products or the associated aesthetics, but is an integral aspect of a sustainable economy, ethical lifestyles and the shaping of communities. The inclusion of opportunities to engage in creative processes within a child's experiences provides a clear narrative that children should use their creative ability to have ownership and take action over their environment based upon Dewey's recognition of learning 'through' and 'by doing'. Such empowerment ultimately leads to the strengthening and further stabilisation of future societies where each individual has the enhanced capability of shaping their destiny. Thus, creativity must be regarded as being much more than transforming or adding value to products. Rather, it is an integral necessity; a component of a sustainable economy and a key determinant in the shaping of future individuals and their societies. As Chomsky (2000) noted, "Citizenship and creativity leads to liberty and enlightenment. It offers new ways of radical thinking, empowerment, autonomy, future shaping, and the challenging of static hegemonies" (p. 38).

Therefore, an initial standpoint is that creativity is an important part of 'being' and, while often associated with the arts, it is also about a broader sense of citizenship and empowerment. Its relationship with design is also often considered symbiotic in that you can't have one without the other. However, I would argue that all design should be creative but not all creativity involves design. I would also propose that while I believe that all design should entail creativity this is not always the case, as often what is perceived as creativity is in fact not necessarily so. Therefore central to both creativity and design, as part of an educational experience, has to be a strong sense of values and ethics.

The ambiguities between design and creativity lie in their perceived symbiosis—evident, for example, in Gardner's (1997) description of creativity as being the ability to solve problems, fashion products and raise new questions, which would seem an equally pertinent definition for design within technology education. Amabile (1990) suggests that a product or response will be judged to be creative to the extent that it is considered novel, appropriate, useful, and correct or valuable in the context of the task in hand. She sees creativity being expressed in situations where domain-relevant skills, creativity-relevant skills and task motivation interact.

Both Gardner's and Amabile's established rationales could be considered as definitions of design activities found in Western-orientated education systems, as they both recognise creative capability through the fashioning of product responses by consideration of 'what might be' rather than 'what is'. This overlap in definitions is understandable if we consider all design to involve creativity. However, my belief is that creativity does not always entail problem solving (a term that is itself open to scrutiny) and does not always involve domain-specific knowledge. As such, creativity is an integral part of design but design is not integral to creativity.

If this is the case then the promotion of creativity is central to design activities. However, a tension often exists between the rhetoric and reality, and claims that assume creativity is an inherent part of design (as part of technology education) are open to scrutiny. The reality is, there is a significant need for creativity to be nurtured and encouraged within design rather than stifled by over-prescription and pragmatism. This has been where 'Creativity in Crisis' (Barlex 2003; Kimbell 2000; Spendlove 2005) exists as a prevailing theme in technology education, highlighting that the subject often fails to offer creative opportunities (Atkinson 2002; Spendlove 2005, 2007a, b) with the conclusion that many of the activities within technology education are not conducive to creativity (Barlex 2003).

## *Barriers to Creativity*

Inevitably assessment remains an issue when discussing creativity in the context of a formal education setting, as often the form of assessment can distort activities towards what can be measured. Keirl (2004) contextualises these concerns, calling for a "culture of creativity" but emphasising that "it is much easier to facilitate a culture of risk taking, questioning and being different if such behaviours are both valued and well managed" (p. 155). The problem remains that politicians, parents and teachers often have difficulty conceptualising creativity in that it is inherently ill disciplined, difficult to manage, difficult to measure and difficult to understand, especially if you consider yourself as being not particularly creative.

Creativity requires a leap of faith, it has to be risky and the returns from it can be low as well as high. Therefore, teachers who are highly accountable, whose reputation and performance are measured through the perceived success of their students' assessed performance, will often—despite their best intentions—provide their students with a benign and impoverished creative experience. Such constraining of creative opportunities in learning experiences can lead to oppressive experiences where students are conditioned into a response necessary for meeting a notionally 'correct' view of predetermined and reorganised knowledge consumption. This *modus operandi* has increasingly dominated much of teachers' pedagogic practice, when one of our core goals as educators should be to maximise the potential for students to be creative and successful learners.

Assessment, however, only represents one of the many potential 'barriers to creativity'. A technology experience can be constrained by various other attributes, such as personality (Amabile 1996), reward, criticism (Berger and Ferguson 1990), and environmental stimulants (Amabile 1988) including time pressure, evaluation, status quo and political problems. In my own research (Spendlove and Wyse 2007) three clear categories of barriers to creativity emerged relating to statutory, organisational and pedagogical factors. Barriers attributed to statutory requirements included statutory tests with an emphasis on 'attainment rather than achievement' and the existence of an 'audit culture'. Organisational barriers were associated with competing demands that existed in schools and difficulties finding time and space (including physical space) to allow children opportunities to be creative.

Pedagogical barriers to creativity include the concept of teachers 'playing safe', resulting in a benign experience for the learner (Spendlove and Hopper 2005; Spendlove and Rutland 2007; Spendlove and Wells 2013)—although teachers wanted to take risks, the effects of accountability were felt to constrain their practice. This is reinforced by Atkinson (2000) who discovered that higher order thinking, such as creativity, problem solving and analytical thinking, were not always capitalised on and could in fact be detrimental for students' 'achievement' because of the need for high levels of performance at public examinations—which fail to reward creativity. A sad indictment of both the subject and the system is illustrated by evidence that creative capabilities are not necessarily required in measured performance and that being highly creative could in fact be a hindrance in terms of examination grading.

Davies (1999) revealed similar contributory factors as barriers to creativity, identifying difficulties that teachers found in promoting creativity through technology education: Teachers were sometimes anxious about their understanding of creativity and frustrated by difficulties in keeping their knowledge and skills updated. Davies concluded that teachers may ultimately be impeding creativity in their students, particularly if they themselves lacked confidence in their understanding of creativity—and were unwilling to take risks due to the legislative and institutional framework in which they operated.

## Doing Design—Designing (or Not)

The act of design—'designing'—is fundamental to technology education. However, the processes for designing and the pedagogical strategies for developing design capability remain problematic when considered in an education setting. Central to this issue is the tension that exists between validity, progression, performativity, manageability and accountability. But before these areas are explored further it is worth reconciling a further tension, which relates to the critiquing of the celebrated 'image' of the designer, since technology education should not be seen to be generating misconceptions as to what designers do.

Very few designers have the freedom to design what they wish in the way that they wish, and as such designing as a profession needs to be seen in its entirety and not be represented by unrealistic 'celebrity' and elite designers who are allowed to be as creative as they wish. Designers often have to appeal to mass markets, on low margins, with short time frames for designing in a highly competitive and consumption-orientated economy, and they are often frustrated by such constraints. Designers are also responsible (at least partly) for over-consumption, obsolescence, damage to the environment, economic vulnerabilities, and deaths through poorly designed objects or through the design of military products, or even the everyday car or motorcycle. Consequently design, as a feature of technology education, should embody a strong ethical component that questions both the profession and process of designing.

Creating a technology education system predominantly seeking to serve the design and technology industries is prone to failure simply because so few children will actually progress into these careers. However, every child will potentially become a powerful consumer, citizen and voter and therefore at the very least should have a broader understanding of key design principles in order to make important judgements in later life (see Chap. 8, this volume, for four different justifications for technology education). As a consequence if we want to educate all children to participate in a complex and evolving society as proactive and empowered citizens then perhaps there should be a considered approach to both value and challenge the status of the designer.

### *What Can We Learn from Designers?*

Having examined the image of the designer, it is important to acknowledge that an enormous amount can be learned from designers—but that it is essential to be selective in what information is used. Therefore, while drawing on some of the key characteristics and strategies that 'successful' designers employ, the recognition that many designers simply do not have such freedom needs to be an accompanying message. Many designers wish to be creative but often have to be risk-averse in order to supply the market and meet the economic demands placed upon them. It is therefore an unfortunate reality that cheap, mass-produced, futile, environmentally damaging, exploitative products have all been 'designed'.

What we do know from examining designers' work is that four major themes have emerged that capture the essence of designer activity, namely, an absence of algorithms, the interaction between problem and solution, a focus on object worlds, and design as a social process (McCormick 1994). If we add to this ethical and sustainable aspects then these five themes provide a useful framework for a progressive technology experience in education.

In many ways, design as an activity within a professional context remains far from clear and as a consequence the process of looking to design-related professions for guidance within an educational context needs to be a continual one, while adopting a critically reflective approach. For example, the ways designers work and are educated, the skills they need, the contexts they operate in, and the nature of their thinking are constantly being redefined. As such the disciplinary, conceptual, theoretical, and methodological boundaries that represent the ever-wider activities and practice (Rogers 2013) of designerly (Cross 2006) activities are being challenged. While 'designerly' thinking provides a valuable resource for 'education', the application in an educational context must be scrutinised.

As indicated earlier many of the difficulties of design activities within technology education are associated with issues of validity, progression, performativity, manageability and accountability. This has most notably been encapsulated by trying to capture, assess and validate student design capability without it being distorted through ritualistic routines—which Atkinson (2000) has identified as perversely militating against the development of higher order thinking skills associated with designing. Unfortunately the ritualisation of designing—largely though the use of 'design portfolios' in order to meet assessment needs—has been seen to distort genuine engagement into a series of contrived and compartmentalised entities rather than a coherent whole (Kimbell 2002; McCormick and Davidson 1996; Spendlove and Hopper 2006; Stables and Kimbell 2000; Welch and Barlex 2004).

Welch and Barlex (2004; Welch et al. 2005) have shown that a dichotomy exists between school practice and professional practice in that while attempting to achieve the same procedural elements, the designers' perception of the portfolio was conceived as representing their best endeavours through 'showcasing' their work. Within a school context, the portfolio represents the steps to achievement. The dichotomy is further illustrated by the performativity of the child acting as a designer in an educational context, which begs the question: What is the role of the teacher in such spaces if the learner is acting as the designer? In many ways this is the crux of the problem in technology education in that much practice is trying to emulate industrial and professional practices, with the learner often adopting the role of apprentice. While such practice may be justifiable and sustainable in an industrial context it is less so in an educational setting when learning is sacrificed in pursuit of performative measures and production of artefacts. In such circumstances the 'product' of the activity becomes 'collusion' and 'coercion' between the various interested parties, such as the teacher, student, parents, and school leaders and administrators.

Such performative distortion is often camouflaged through a manufactured product endorsed through a system of apparent success. The collusion is perpetuated by the accountability, which comes from the pressure to achieve for both

**Table 9.1** Learning scenarios

| Scenario | Paradox of learning |
|---|---|
| The student 'knows' the designed product is good and the teacher 'knows' the designed product is good | This may mean the designed product is genuinely 'good' or that both the teacher and student view 'good' as meeting the performative measures required while failing to understand and/or acknowledge the broader context of design activity beyond the performance system |
| The student 'knows' the designed product is bad but represents good measured performance and the teacher 'knows' the designed product is good | Examples of this occur where a student understands the game being played and where the student may have a more informed perspective on design than the teacher/performance system but is willing to subvert their aspirations in order to meet the required performance measures |
| The student 'knows' the designed product is good but the teacher 'knows' the designed product is bad but represents good measured performance | In this context the driver is the performative measure, which distorts the teaching leading to a 'critical schizophrenia' where the teacher coerces the student, creating the illusion of success |
| The student 'knows' the designed product is bad but represents good measured performance and the teacher 'knows' the designed product is bad but represents good measured performance | Such contexts represent compliance by both the teacher and student, who both recognise the performative measures distorting the designed product but are willing to trade such acknowledgement in pursuit of the illusion of success |

student and teacher by measured performance in terms of exam accountability and notional 'academic' success. Zizek (2001) represents this as students (and teachers) having a notional 'forced choice', behaving as if they had free choice of procedures when in fact their choice is benign. Therefore we can view this paradox of learning as a coerced activity through a series of basic scenarios See Table 9.1, for example.

The point of highlighting the above scenarios is that there exists a danger in technology education of distorting opportunities away from rich and challenging activities towards a formulaic, contrived activity that gives the appearance of genuine learning. In such circumstances of contrived activity, the real 'product' is the sense of collusion and illusion that takes place between the teacher and the student while the material product and learning opportunities become of lower priority. The rationale for such an approach represents the safety and security of being able to predict and micro-manage a 'creative' and 'learning' activity (the equivalent of putting stabilisers onto a child's bicycle but never removing them) and as such the process becomes expedient and illusionary. Consequently the 'process' becomes a vehicle for 'learning' in its own right and "the artificial 'design process' as an educational vehicle is thus born" (Liddament 1996, p. 1).

The above does not suggest that designing doesn't or can't take place. Rather, it confirms that designing cannot exist without creativity, riskiness and uncertainty—three characteristics that have become marginalised as teachers and pupils become more accountable in terms of productivity and performativity. The significance of this is the translation into pedagogic practice where teaching becomes instructional, focusing on leading learners through a series of artificial and contrived steps. In such circumstances, undemanding or preconceived tasks can give the appearance of students designing and progressing when the reality is learners who lack autonomy or creative opportunity. The unfortunate consequence is that rather than teachers challenging many of the misconceptions that surround the concepts of 'designing', they in fact contribute to such misinterpretation. Inevitably this generates the question of what technology education is for, as the danger in adopting dysfunctional pedagogical models is that learners neither engage with the broader aims of the subject *or* the technical, academic or vocational opportunities also afforded.

New ways of capturing and assessing authentic processes have however been generated (Kimbell et al. 2009; Pagram and Williams 2010, see also see Chap. 7, this volume) which offer some hope that many of the manageability and accountability issues in facilitating a high quality design experience can be overcome. Therefore, while some of the discussion above may appear a pessimistic view of design as a part of technology education, it is countered by the enormous opportunities that many inspirational teachers offer their students. These are where teachers have a broader understanding of design, problem finding, creative learning, and technology education, and a clear view of what they wish to achieve in their environment and how it will benefit their students beyond performance in assessed outcomes. In such environments teachers recognise the need to engage learners in a rigorous, creative, designerly, authentic experience that is driven by learning, while also meeting institutional requirements in a meaningful rather than contrived way.

In many ways this is the norm of good creative practice, where the approach adopted by the teacher is inherently risky in that the teacher models the very attributes they wish to develop in their learners. It is also what I would hope would be considered the norm of practice, and I now move on to consider what I feel offers even greater opportunities for the inclusion of design, both as part of technology education and general education.

## Design Thinking

Victor Papanek (1971) in his seminal work *Design for the Real World* wrote, "Design is a conscious and intuitive effort to impose meaningful order" (p. 4). This quote highlights a position that many in technology education occupy. However, I feel it is worth reconsidering the assumption that design is a conscious and intuitive act, particularly in relation to 'design thinking', which is often regarded as a methodology to generate innovative ideas (Brown 2009). As Rogers (2013) points out, there is no "universally agreed upon definition of 'design thinking', but the strongest common denominator embraces the centrality of the user and empathy to the human condition" (p. 434).

My starting point for exploration is therefore that designerly decision-making—'thinking' within a design context—while being a human condition, is vulnerable to the vagaries of human cognition. As such design thinking is a subconscious activity with the appearance of consciousness that utilises often defective decision-making process. It would therefore appear that any form of education for design capability (presumably within technology education) should involve insights into the limitations of designerly thinking and the inherent 'cognitive traps'. While such 'flawed heuristics' may be perceived as intuitive and rational, we have to contest such thinking in a learning environment based on the acknowledgement of the existence of cognitive limitations and irrational thinking.

As a consequence of the irrationalities we all have, exploration of such phenomena provides unique opportunities for technology education. Examining designerly thinking within a context of 'imposing meaningful order' while attempting to improve the world we live in as part of technology education seems a unique and valuable educational opportunity. Central to this is exploring the concept of 'agency' and whether humans have the ability to operate autonomously without preconceived ideas driven in the pursuit of pure self-interest. Within this context 'agency' should be considered as the socio-culturally mediated capacity to act (Ahearn 2001) within the constraints of the psychological determinants of human functioning. Such functioning includes the desire and efficacy beliefs (Bandura 2006) that we can produce effects from our actions, and that without such beliefs there is "little incentive to act, or to persevere in the face of difficulties" (p. 170). Therefore 'agency' contributes significantly to motivation, emotional well-being, and performance accomplishments—but is itself socially, culturally and psychologically shaped and constrained.

Related to the concept of a lack of agency is a well known fable about a scorpion that wants to cross a river but cannot swim. The scorpion convinces a frog to carry him across and promises not to kill the frog: "If I try to kill you, then I would die too, for you see I cannot swim!" The frog takes the scorpion across the river and halfway across the frog suddenly feels a sharp sting in his back and out of the corner of his eye, sees the scorpion remove his stinger from his back. "You fool!" croaked the frog, "Now we shall both die! Why on earth did you do that?" The scorpion shrugged "I could not help myself. It is my nature."

This fable neatly caricatures a potential lack of agency humans have, which is best illustrated through our 'conscious' destruction of the planet through over consumption and living beyond our means with inevitable consequences. The relevance of this is that we as humans are designers of our destiny and do have some sense of agency, albeit socially, psychologically and culturally constrained. However, it would appear we are all susceptible to elements of irrationality—just like the scorpion, it is in our nature. This is illustrated by the self destruction observed through environmental damage, obesity, terrorism, obsolescence, war, etc. However, such irrationality isn't confined to these overt concepts of self destruction as it is prevalent in everyday decision-making—which I believe all those in design decision-making (or not) need to be aware of.

## *What Does This Mean for Me as a Teacher?*

Perhaps no doubt creeping through your mind is the question, *What has this got to do with my classroom and the teaching of design as a part of technology education?* My answer is "everything". As a species we have a perceived intelligence that is sufficiently limited to fully recognise or innately understand our own limitations. However, when making decisions through the altruistic act of designing to improve the quality of others lives by 'imposing meaningful order' we need to challenge learners through the lens that acknowledges that our cognitive apparatus and decision-making is often flawed. Such a task is not easy, yet seems an essential part in any future-focused learning activity that seeks to mediate and shape the world we live in, particularly in an educational context.

In previous publications (Spendlove 2007a, b, 2008) I have discussed the role of emotions and sub-conscious processing and in particular how they influence our decision-making. For example, our conscious awareness of stimuli from the environment lags behind actual perception by approximately half a second. Therefore a backward referral of subjective experience results in an individual's perception of the stimulus and its conscious awareness as simultaneous. Such realities can be slightly destabilising, particularly when we base our existence on what we believe are realities—when in fact our perceptions of reality may not always be what they appear to be. Equally we are potentially only 'receiving' 1/200,000 of what is actually being processed by our senses and as such our perceived realities are merely highly filtered experiences. Interestingly, perceptual awareness varies to some extent by gender, genetic and cultural group. Therefore the judgements we make are often constrained and unique to a particular group that we may be part of, and such distinctive processing influences our priorities and decision-making subconsciously. In other words, what we may think as straightforward and consistent with others is not necessarily so and what we see, hear, feel and believe is often a distorted version of reality.

Approximately 85 % of the time, analytical decisions that we think we are making have already been 'primed' by the unconscious mind while at the same time we are innately looking for patterns that may not exist. Such extreme 'apophenia' is recognised as the propensity to spontaneously perceive connections and meaningfulness in unrelated phenomena (Carroll 2003). In many ways this is ultimately a creative active act, as in making unusual connections, but it is also a psychotic act in that meaningful connections between unrelated events are made. Relativity, imprinting, anchoring and arbitrary coherence are all further cognitive limitations that add to our consistently illogical processing, particularly when reasoning under various degrees of uncertainty—which is the position that any design must operate in.

In many ways the overriding antidote to avoid engagement with the cognitive limitations is to rely on 'intuition' and 'gut feeling'. Indeed, we are often steered by such feelings. However what we feel about something informs what we think and not the other way around (Damasio 2006). As such intuition or deliberate practice is when we operate "quasi-automatically and with reasonable proficiency" (Pigliucci 2012, p. 103) and is an instinctive sense that something is right, a heuristic shortcut, but which again is prone to errors when engaged in decision-making. A further way of overcoming our limitations is through the adoption of purely optimist strategies. 'Optimism bias' is a key survival strategy as we mentally project forward and identify our future needs. Again, this is prone to error, which Sharot (2012) cites as the "superiority illusion" in that we tend to think we are better than we are. Sharot gives an example of a survey of driving where 93 % of respondents indicated they were above average in driving ability, which is statistically impossible.

While identifying the many limitations that exist in terms of our ability to make design decisions in the interest of others, there is also the other side—the exploitation of the limitations in others by designers. Kahneman (2011) identifies a "focusing illusion" where we misjudge the potential impact of certain circumstances, such as not considering the unintended consequences of a product's misuse. In addition, designers and marketing specialists also exploit consumer demand by offering a 'better future' using a combination of focusing illusion and visual illusions while capitalising on an optimism bias, manipulating consumer emotions and thriving on the many cognitive limitations identified above.

As previously indicated, it appears that any form of education for design capability should develop an insight into the limitations of designerly thinking by acknowledging our inherent cognitive flaws. Therefore when considering improving the lives of others we need to recognise how these predictably irrational traits interplay with a whole range of cognitive, visual, emotional, evolutionary, narcissistic and cultural pitfalls which ultimately limit and distort our thinking.

## Conclusions

Dewey (1916) is quoted as saying that "nothing has brought pedagogical theory into greater disrepute than the belief that it is identified with handing out to teachers recipes and models to be followed in teaching" (p. 107). If we fail to acknowledge this mantra in technology education by delivering design as a highly prescribed, sterilised and artificial learning experience then we will fail to engage learners with the challenges inherent in an uncertain future.

In this chapter, I have attempted to be deliberately provocative and offer a slightly different perspective on how design might be viewed as a reflective and metacognitive activity that acknowledges and explores the frailties of the human cognition within a context of technology education. There is a significant amount of additional literature about strategies, pedagogy and practice within this book that I hope the reader will engage with while also holding in mind some of what has been offered in this chapter. Much of what is presented here is, however, recognised as being challenging and no easy solutions are offered.

There is a need to transcend the dichotomy of design being both a verb and a noun and move to a position that recognises design in different situations and contexts as being different. This means recognising design as part of technology in an education context as different to design in other contexts. It also means recognising design education at the different phases of general education as different entities with different objectives and outcomes. In doing so, there is a need to recognise and clarify the association with vocational aspects of design. This means acknowledging that design as an educational activity is undoubtedly linked to the associated professional fields of practice related to design. However, it is important to explicate the essential educational features that may be of value.

The recurring theme within this chapter is therefore the need to distinguish between design as a professional activity (which in itself is an unstable and evolving context) and design as an educational activity. By identifying this, I have also attempted to distinguish between design as an activity and design as a process. Roberts et al. (1992) also made this distinction by identifying that:

> [designing] consists in relating and drawing attention to purpose, the self, and the means and significance of man's [sic] intervention in his habitat
>
> [while design as an activity] distinguished from design-educational activity, is directed towards the manipulation of things and systems so as to achieve the most acceptable and practicable fit between a particular set of desires and needs, on the one hand, and a particular means for fulfilling them, on the other. (p. 3)

Paradoxically, Roberts et al. also indicated that design activity (as in, a professional activity) is "concerned more with the attainment of a result than with the acquisition of knowledge" (ibid.). However, the changes in education in the last 20 years suggest that, increasingly, educational attainment in design-related activities has reversed this relationship, with design in general education increasingly focused on attainment (of a result), leading to a distortion of design activities.

The paradox above is unfortunate as design activities provide the opportunity to engage in rich learning "to do with self, self in relation to made things and systems, and the appreciation of the effect of his or her own, and other people's activity in and on the world" (Roberts et al. 1992, p. 3). As such, in many educational design activities we have the equivalent of 'swimming without getting wet' in that what can be measured, in the form of attainment, distorts what should be inherent and valued. As a consequence, we often don't engage with the risky elements central to good practice. This confirms the adage that we tend to 'value what we can measure rather than measure what we value'. As a consequence, a sterile form of design education can exist that fails to recognise how technology and design have interacted within a political and free-market economy to shape our contemporary culture. A design education bereft of an understanding of social, psychological, philosophical, political, economic and cultural influences appears to be incredibly limited.

Ultimately, we can see how the socio-cultural, neurological, psychological and emotional illusions influence our everyday epistemological and ontological beliefs, which are encapsulated in the decisions that we make in attempting to design a world for others. Such intent in design is an essential human activity that allows us to be empowered to change the world. Often, however, the message in education distorts this empowerment—instead of enlightening users to have a voice, this voice is suppressed through reducing ownership of the design activity. Therefore, opportunity for design enlightenment becomes a disenfranchising activity.

If design cannot take place within a technological learning activity then perhaps it should occupy a different space within the curriculum, as it seems too important to marginalise. Current educational constructs, far from challenging epistemological illusions, reinforce a whole series of fallacies and the reproduction of knowledge in the form of positivist Newtonian linearity (Williams and Arrigo 2002) that has become the *de facto* operating system of society. I have argued, therefore, that alternative 'future' models of technology education can offer a productive means to enhancing education and children's understanding of designing as opposed to current models that merely proliferate the illusion of progression and understanding.

In offering a conclusion I am conscious (I think) that my own cognitive limitations, cultural assumptions and emotional influences have shaped my thinking. However, what I have presented does not imply a form of determinism as the 'escape hatch' (Stanovich 2004) is through recognition of such human limitations through engaging with technology education and the shaping of a future world. In recognising these as a starting point, while acknowledging such constraints, we also recognise that the potential damage from imposing moderated beliefs should not be underestimated. Therefore a future curriculum predicated by learning to challenge assumptions in reaction to the excesses of consumption provides a means for challenging the automatic reproduction of our beliefs, cognitive limitations, cultural influences and emotional manipulation upon notional autonomous decision-making. Such a curriculum also challenges the rational altruistic status of the designer and is central to the needs of technology education.

Pragmatists may scratch their heads in the consideration of such propositions, yet the significance of 'designing', as an activity, in attempting to understand the world through recognition of one's own limitations, shouldn't be underestimated. Starting with a deficit model of vulnerable thinking and challenging the boundaries of conventional thinking should be central to the aim of education. As a consequence, teachers should not seek to ease learner transitions across boundaries, through promotions of illusions, as the engagement of 'pedagogies of discomfort' is of greater value than the illusions of accomplishment that are proliferated through current models of achievement. As such, technology education is uniquely placed to both be explored and be used to explore possibilities as we seek to create a sustainable future.

## References

Ahearn, L. M. (2001). Language and agency. *Annual Review of Anthropology, 30*, 109–137.

Amabile, T. M. (1988). A model of creativity and innovation in organizations. *Research in Organizational Behaviour, 10*, 123–167.

Amabile, T. (1990). Within you, without you: The social psychology of creativity, and beyond. In M. Runco & R. Albert (Eds.), *Theories of creativity* (pp. 61–91). London: Sage.

Amabile, T. (1996). *Creativity in context: Update to the social psychology of creativity*. Colorado: West View.

Atkinson, S. (2000). Does the need for high levels of performance curtail the development of creativity in Design and Technology project work? *International Journal of Technology and Design Education, 10*(3), 255–281.

Atkinson, S. (2002). Creativity versus the need for high levels of performance. In G. A. Owen-Jackson (Ed.), *Teaching Design and Technology in secondary schools* (pp. 161–176). London: Routledge Taylor Francis.

Bandura, A. (2006). Toward a psychology of human agency. *Perspectives on Psychological Science, 1*, 164–180.

Barlex, D. (2003). *Creativity in crisis? Design and Technology at KS3 and KS4* (DATA research paper 18). Wellesbourne: DATA.

Berger, F., & Ferguson, D. H. (1990). *Innovation: Creativity techniques for hospitality managers.* New York: Wiley.

Brown, T. (2009). *Change by design: How design thinking transforms organizations and inspires innovation.* New York: Harper Business.

Carroll, R. T. (2003). *The skeptic's dictionary: A collection of strange beliefs, amusing deceptions, and dangerous delusions.* Hoboken: Wiley.

Chomsky, N. (2000). *Chomsky on MisEducation* (D. Macedo, Ed.). New York: Rowan and Littlefield.

Cross, N. (2006). *Designerly ways of knowing.* London: Springer.

Damasio, A. (2006). *Descartes' error: Emotion, reason and the human brain.* London: Vintage.

Davies, T. (1999). Taking risks as a feature of creativity in the teaching and learning of design and technology. *The Journal of Design and Technology Education, 4*(2), 101–108.

Dewey, J. (1916). *Democracy and education.* New York: The Free Press.

Flusser, V., & Cullars, J. (1995). On the word design: An etymological essay. *Design Issues, 11*(3), 50–53.

Gardner, H. (1997). *Extraordinary minds.* New York: Harper Collins.

Heskett, J. (2001). Past, present, and future in design for industry. *Design Issues, 17*(1), 18–26.

Kahneman, D. (2011). *Thinking, fast and slow.* New York: Farrar, Straus and Giroux.

Keirl, S. (2004). Creativity, innovation and life in the lily pond: Nurturing the design and technology family while keeping the alligators fed. *The Journal of Design and Technology Education, 9*(3), 145–157.

Kimbell, R. (2000). *Critical concepts underpinning the Design & Technology curriculum in England.* Keynote address, international Technology Education conference, University of Brunswick, Germany.

Kimbell, R. (2002). Assessing design innovation: The famous five and the terrible two. In E. W. L. Norman (Ed.), *DATA international research conference 2002*. Wellesbourne: The Design and Technology Association.

Kimbell, R., Wheeler, T., Stables, K., Shepard, T., Martin, F., Davies, D., Pollitt, A., & Whitehouse, G. (2009). *e-Scape portfolio assessment: A research & development project for the Department of Children, Families and Schools. Phase 3 report.* London: TERU, Goldsmiths, University of London.

Liddament, T. (1996). Design and problem solving. In J. S. Smith (Ed.) *International Design and Technology Educational and Research Conference (Idater 96),* Loughborough University, UK.

Love, T. (2002). Constructing a coherent body of theory about designing and designs: Some philosophical issues. *Design Studies, 23*(3), 345–361.

Mau, B. (2007). "What is massive change?" *Massive change.* Accessed 15 Jan 2012 from: http://www.massivechange.com/about/

McCormick, R. (1994). Learning through apprenticeship. In D. Blandow & M. J. Dyrenfurth (Eds.), *Technology education in school and industry* (pp. 16–36). Dordrecht: Kluwer.

McCormick, R., & Davidson, M. (1996). Problem solving and the tyranny of product outcomes. *Journal of Design and Technology Education, 1*(3), 230–241.

Pagram, J., & Williams, P. J. (2010). Digital assessment: A Western Australian experience in a senior secondary engineering course. In D. Spendlove & K. Stables (Eds.), *Ideas worth sharing. The Design and Technology Association Education & International Research Conference, Keele University 2010* (pp. 85–92). Wellesbourne: Design and Technology Association.

Papanek, V. J. (1971). *Design for the real world: Human ecology and social change*. Chicago: Thames & Hudson.

Pigliucci, M. (2012). *Answers for Aristotle: How science and philosophy can lead us to a more meaningful life*. New York: Basic Books.

Roberts, P., Archer, B., & Baynes, K. (1992). *Modelling: The language of designing* (Design: Occasional paper no. 1). Loughborough: Loughborough University.

Rogers, P. (2013). Articulating design thinking. *Design Studies, 34*, 433–436.

Sharot, T. (2012). *The optimism bias: A tour of the irrationally positive brain*. New York: Vintage.

Spendlove, D. (2005). Creativity in education: A review. *Design and Technology Education: An International Journal, 10*(2), 9–18.

Spendlove, D. (2007a). *We feel therefore we learn: The location of emotion in a creative and learning orientated experience*. Keynote paper presented at the Design and Technology Educational and International Research conference, Wolverhampton.

Spendlove, D. (2007b). A conceptualisation of emotion within Art and Design Education: A creative, learning and product orientated triadic schema. *International Journal of Art and Design Education, 26*(2), 155–166.

Spendlove, D. (2008). The locating of emotion within a creative, learning and product orientated Design and Technology experience: Person, process, product. *International Journal of Technology and Design Education, 18*, 45–57.

Spendlove, D., & Hopper, M. (2005). *Coping with uncertainty: an exploration of the place of emotion within the creative curriculum*. Paper presented at the British Education Research Association annual conference, Glamorgan.

Spendlove, D., & Hopper, M. (2006). Using 'electronic portfolios' to challenge current orthodoxies in the presentation of an Initial Teacher Training Design and Technology activity. *International Journal of Technology and Design Education, 16*, 177–191.

Spendlove, D., & Rutland, M. (2007). Creativity in Design and Technology. In D. Barlex (Ed.), *Design and Technology—for the next generation* (pp. 140–153). Shropshire: Cliffeco.

Spendlove, D., & Wells, A. (2013). Creativity for a new generation. In G. Owen-Jackson (Ed.), *Debates in design and technology education* (pp. 166–179). Oxon: Routledge.

Spendlove, D., & Wyse, D. (2007). Creative learning: Definitions and barriers. In A. Craft, T. Cremin, & P. Burnard (Eds.), *Creative learning 3–11 and how we document it: What how & why?* (pp. 11–18). Stoke on Trent: Trentham Books.

Stables, K., & Kimbell, R. (2000). The unpickled portfolio: Pioneering performance assessment in design and technology. In R. Kimbell (Ed.), *Design and technology international millennium conference* (pp. 195–202). Wellesbourne: DATA.

Stanovich, K. E. (2004). *The robot's rebellion: Finding meaning in the age of Darwin*. Chicago: University of Chicago.

Welch, M., & Barlex, D. (2004). Portfolios in design and technology education: Investigating differing views. In E. W. L. Norman, D. Spendlove, P. Grover, & A. Mitchell (Eds.), *DATA international research conference 2004* (pp. 193–197). Sheffield: Sheffield Hallam University.

Welch, M., Barlex, D., & Taylor, K. (2005). I don't enjoy making the folder: Secondary students' views of portfolios in technology education. In E. W. L. Norman, D. Spendlove, P. Grover, & A. Mitchell (Eds.), *DATA international research conference 2005* (pp. 175–180). Sheffield: Sheffield Hallam University.

Williams, C. R., & Arrigo, B. A. (2002). Law, psychology, and the 'new sciences': Rethinking mental illness and dangerousness. *International Journal of Offender Therapy and Comparative Criminology, 46*(1), 6–29.

Zizek, S. (2001). What can Lenin tell us about freedom today? *Rethinking Marxism, 13*(2), 1–9.

# Chapter 10
# The Alignment of Technology with Other School Subjects

**Cathy Buntting and Alister Jones**

This chapter examines the traditional separation, or siloing, of knowledge domains into distinct school subjects—languages, mathematics, science, social science, the arts, and technology—and considers the benefits and challenges of breaking down these silos and bringing about closer alignment between technology as a school subject, and other subject areas. One key purpose for aligning technology with other subjects is to increase the scope and opportunities for students to develop relevant skills and dispositions to address 'wicked problems'—complex problems with multiple causes and interdependencies that are difficult or even impossible to solve, or even define, using the tools and techniques of only one organisation or discipline.

## Introduction

In a world of unprecedented social, economic and technological changes and an exponential increase in human knowledge, the role and purpose of education needs to be examined and re-visioned. How do we 'do' schooling in a context defined by change and increasing complexity, fluidity and uncertainty? What skills will be needed to solve the world's complex problems, and how do we support students to develop them? These are crucial questions that need to be considered if we are to explore the future of technology education in any meaningful way.

According to the '21st century' or 'future-focused' education literature (e.g., Broadfoot 2000; Brown 2006; Gilbert 2005; Young 2010), there is an urgent need

C. Buntting (✉)
Faculty of Education, University of Waikato, Hamilton, New Zealand
e-mail: buntting@waikato.ac.nz

A. Jones
Deputy Vice-Chancellor, University of Waikato, Hamilton, New Zealand
e-mail: a.jones@waikato.ac.nz

 
PJ. Williams et al. (eds.), *The Future of Technology Education*, Contemporary Issues in Technology Education, DOI 10.1007/978-981-287-170-1_10

to think differently about what schools are for, what students should learn in them, and how 'success' might be measured:

> 'Traditional' forms of education, it is argued, were designed to develop knowledge and skills valued in 20th century social and economic conditions, and are no longer appropriate in the 21st century environment. New approaches are needed if our young people are to develop the 'dispositions' (to knowledge, thinking, learning and work) needed to productively engage in the 21st century world. (Bull and Gilbert 2012, p. 4)

This chapter examines the traditional separation or siloing of knowledge domains into distinct school subjects—languages, mathematics, science, social science, the arts, and technology—and considers the benefits and challenges of breaking down these silos and bringing about closer alignment between technology as a school subject, and other subject areas. Importantly, the term 'alignment' adopts here the nuanced definition of 'joining with others in a cause'. In other words, emphasis is placed on the *purpose* of the alignment.

One purpose for aligning technology with other subjects is to increase the scope and opportunities for students to develop relevant skills and dispositions to address 'wicked problems'—complex problems with multiple causes and interdependencies that are difficult or even impossible to solve, or even define, using the tools and techniques of any one organisation or discipline (Bull and Gilbert 2012). Standard examples include climate change, natural hazards, public healthcare and nuclear energy and waste. While it is likely that solutions will involve technological intervention, they will also probably require large numbers of people to change their views and/or behaviours. In addition, efforts to solve one aspect will tend to reveal or create other problems. Multidisciplinary, system-focused approaches will therefore be needed. What might technology education look like, and how might it be taught, if it is to prepare students to contribute to these types of solutions?

## Traditional Subject Silos

Technology is a relatively new subject within school curricula around the world, representing a broadening of more traditional, gendered crafts and skills-based subjects to include aspects of design and/or notions of technological literacy for all students. However, unlike school subjects with a far more established history—science or mathematics, for example—there are large differences in where the emphasis is placed in the technology curricula of different countries. For instance, Jones et al. (2013) identify seven representations of technology education: skills and gendered craft subjects; industrial arts and/or vocational training; technology informed by design; technology as applied science; technology integrated within Science, Technology, Engineering and Mathematics (STEM); multiple technologies (process technologies, manufacturing technologies, agritechnologies, biotechnologies, etc.); and technological literacy. In the majority of countries—and sometimes in provinces or states within countries—a range of these representations is offered, particularly at different levels of schooling. Further, while many jurisdictions have moved towards notions of technological literacy in their rhetoric and guidelines, this is often not reflected in schooling practice.

Technology's position as a relative newcomer in school curricula and its ongoing development is both a strength and a weakness (Jones et al. 2013). As a strength, technology education can build from and reflect on the development of other education disciplines in terms of developing curricula, pedagogy, and assessment approaches. Weaknesses lie in its fragility in terms of its status as a subject and the socio-political environment of schooling, the need to continue to clarify important concepts and relevant pedagogies, and the support required for teacher preparation and professional learning in relation to a relatively new subject area.

Unfortunately, where technology has been formally aligned with other subjects, such as science, it has tended to be marginalised. This problematises any discussion about cross-disciplinary initiatives, since it is technology's establishment as a separate subject that has granted it inherent status, albeit somewhat fragile. David Barlex, for example, points out in Chap. 8 the need for technology education to have own identity, arguing:

> Relationships between technology and other subjects that are mutually beneficial can only be developed and sustained if technology has a clear sense of self and its own unique contribution to education. Without this, relationships with other subjects become ones in which their educational goals dominate and technology is always the lesser partner, sometimes to the extent that it is no longer true to its unique nature.

In other words, the integrity of the learning goals for technology education can be retained in cross-curricular initiatives but only if technology education has a strong identity that is valued.

## Socio-political Drivers for Cross-Disciplinary Initiatives

Although the interdisciplinary nature of contemporary scientific and technological research is not reflected in the traditional siloing of school subject areas, rhetoric regarding the synergies between technology and science education is not new. For instance, science curricula have included technological applications of scientific concepts for several decades. This approach was driven by recognition of the interrelationship between scientific understanding and technological advancement, spurred on by post-war education and economic policy in the 1940s and 1950s and the USSR's initial forays into space in the 1960s. In the 1970s, the science, technology and society (STS) movement was pioneered (e.g., Gallagher 1971). This made the inclusion of technology into science more overt, although the representation of technology was often as applied science.

More recently, a number of countries have started exploring policy initiatives around science, technology, engineering and mathematics (STEM) and even science, technology, engineering, arts and mathematics (STEAM)—both movements underpinned by political rhetoric related to the importance of science, technology and innovation to economic and social wellbeing. Similarly, 'innovation education' is increasingly being recognised as critical for students' preparation for contributing to their own and their country's future (e.g., Shavinina 2013). Again, scientific *and* technological skills and concepts are recognised as important for students' development in this area, but there is only limited exploration of what skills and/or dispositions might be needed, and how to develop these within current school curricula (Jones and Buntting 2013).

In many of the formerly mentioned initiatives, *science* continues to be privileged with a lesser focus on technology and/or engineering. One reason appears to be associated with the historical privileging of science as an intellectual pursuit with higher social status when compared with the learning of technical and technological skills, which are viewed as being less academic and have traditionally been associated with lower socioeconomic positioning. The result is a valuing of science over technology at the school level as well as by society more generally. Second, it is often left up to science teachers to implement initiatives such as STS and STEM and so the science subculture within which the teacher is embedded influences how these innovations play out in the classroom. In addition, the strength of the teachers' science understanding and skills when compared with their technological knowledge and/or skills often means that science learning predominates. In general, then, while technology is valued in the rhetoric, it is accorded less status in classroom practice and by wider society.

One challenge to developing students' and teachers' technological knowledge and skills—whether in isolation from or together with developing scientific and/or mathematical knowledge and skills—is the ever-widening scope of technological activity, and the different knowledge and skills required in each area of endeavour. For example, the knowledge and work of a mechanical engineer differs substantially from that of a biotechnologist. This highlights the need to think about 'technologies' rather than technology as a single endeavour (Jones 2012). A common response by schools is to package different technologies into discrete subjects, for example, materials technology, food technology, fabric technology, electronics, etc. For a whole host of historical and practical reasons, these subjects are often taught independently of what students might be learning in other subjects, including other technology subjects—except at the elementary/primary level where one teacher tends to be responsible for the teaching of all subject areas. Here, it is conceivably more practical for the teacher to know what is being learnt in each subject area, and to design a programme that maximises transfer of students' ideas and thinking between different parts of their school programme. However, this transfer of learning across subjects relies heavily on the teacher having strong pedagogical content knowledge (PCK) within each domain, including technology.

## Teacher Knowledge

Teacher knowledge is key if technology learning outcomes are to be identified, pursued, and assessed. This of course applies to school programmes where technology is taught as a separate 'subject', but it is perhaps even more important in courses where technology is aligned—or integrated—with other subjects. Consider, for example, a scenario where a technological context is used to underpin learning in one or more other subjects—science, for instance, or social science. If understanding the nature of technology is an important learning outcome, the teacher needs to know how the nature of technology differs from the nature of science or social science.

Even before a teacher can facilitate learning related to specific learning outcomes, appropriate learning outcomes need to be identified. To do this for technology education, the teacher needs to know which aspects of technological understanding or skill development are relevant to the learning context, and be able to make these explicit. This in itself can be difficult, particularly in cases where the teacher comes from a different subject subculture (e.g., a secondary school science teacher or home economics teacher teaching food technology) or where there is a new or revised curriculum with new or changed learning outcomes (as has happened with the international move to include aspects of the nature of technology in technology curricula).

Teacher knowledge also impacts on the ways in which teachers think about how students make progress when learning technology. Decisions about what information to collect and how to interpret it are influenced by the teacher's understanding of the nature of technology, and their orientations to technology teaching and learning. While assessment in technology education is discussed in depth by Kay Stables (Chap. 7, this volume), it is important to note that teacher knowledge of the nature of technology, relevant learning outcomes, and the associated conceptual or procedural knowledge, is paramount. This applies to both summative assessment, and more formative approaches, or assessment for learning (AfL). For this reason, Cowie et al. (2013) extend a conventional definition of AfL to highlight the discipline-specific context:

> Assessment for learning encompasses those everyday classroom practices through which teachers, peers and learners seek/notice, recognise and respond to student learning, throughout the learning, in ways that aim to enhance student learning and student learning capacity and autonomy. *Assessment for learning also needs to reflect, be responsive to, and build on from how particular disciplines generate and legitimize meaning*. (p. 10, original emphasis)

## *The Nature of Technology and Nature of Science*

The alignment of technology with science is probably the most common pairing of technology with another subject area. In particular, movements such as STS, STSE (science, technology, society and the environment) and STEM have offered tantalising opportunities for more integrated approaches to science and technology education. In the majority of cases, however, the emphasis in classroom practice has remained on the science, with the social relevance of technology valued far more highly than concepts associated with the nature of technology. Even the inclusion in science curricula of socio-scientific issues (SSI), where the issue is often associated with a technological outcome of science (e.g., genetic modification of organisms for enhanced crop yield), the scientific knowledge and ethical dimensions tend to be prioritised over technological learning. However, it is sometimes the broader technological context that needs to be considered in order for a full understanding of the ethical issues to be developed. For example, the outcomes that can be achieved in a strictly-controlled laboratory environment often do not reflect the reality of full commercialisation. By ignoring this aspect, the complexity of developing a commercial product is underplayed, obscuring the wider knowledge base required for full product development and perhaps reinforcing a notion of science and technology as being both simple and value-free (Buntting and Jones 2009).

The nature of technology is outlined in a number of chapters in this volume, and comparisons of technology and science can be found elsewhere (e.g., de Vries 2009; Lewis et al. 2007; Williams 2002). They are therefore not the focus of this chapter, suffice to highlight that significant differences do exist. The development of teacher practices have strong links with their initiation and socialisation into particular subject subcultural settings, as already mentioned. This often leads to a consensual view about the nature of the subject, the way it should be taught, the role of the teacher, and what might be expected of students. Thus, science teachers' understandings and perceptions of technology are likely to influence how their students' understandings of technology develop. Similarly, technology teachers' understandings and perceptions of science will influence how their students' understandings of science develop. In the case of the latter, where science is taught through a technological context, it is necessary for teachers to help students make salient the concepts that are relevant to science (Roth et al. 2001).

As an example of the importance of teacher knowledge in identifying and valuing technological learning outcomes, Campbell and Jobling (2012) lament that opportunities to recognise technological knowledge are lost, or at least not maximised, in large international competitions such as the SEAMEO young scientist awards held in Malaysia every second year. While three of the eleven special awards that are made relate specifically to technology, Campbell and Jobling point out that design technology is the driving force in many of the projects, rather than science, and that students' application of technological knowledge and skills could be assessed. Their vision is for technological knowledge to be included as one of the evaluation criteria for future projects. This would require an operational definition of technological knowledge to be established and the students' mentor teachers to help students to recognise and articulate their technological as well as their science learning.

Cowie et al. (2013) illustrate the value of primary teachers identifying specific science *and* technology learning outcomes in their programme planning and implementation. An important reason for identifying and making explicit science and technology learning outcomes—at both the primary and secondary level—is the doorway that this offers to students to identify with science and/or technology and to form identities of themselves as 'scientists' and 'technologists'. Such identity formation, and the development of an interest in and affiliation with technology in both the short term and longer term (Kelly et al. 2008), is surely one of our educational objectives in offering school programmes that use technology as an underpinning context.

At secondary school level, interactions between science and technology departments can be fraught with difficulties. For example, Lewis et al. (2007) reports on a study investigating one British school's effort to foster collaboration between the science and design and technology (D&T) departments. They note how the different philosophical commitments of the science and D&T departments led to antagonistic interactions among colleagues, with each department affirming and prioritising its own learning goals and unable to find areas of sufficient overlap where each subject could helpfully inform students' learning in the other. This led, in the end, to the creation of a 'Science and Technology Week' where both subjects were given special status within the school's activities. Teachers from both departments collaborated together to offer a programme of constructional activities that prioritised students' participation, enjoyment and motivation, and the notion of particular subject requirements was temporarily suspended.

Lewis et al. (2007), discussing the study, suggests that:

> Developing interaction between science and D&T that brings into being a new curriculum will require an exercise in creative collaboration … [where] there is a dynamic integration of expertise achieved through a fluidity of roles fuelled by a common vision and underpinned by trust. (p. 55)

They go on to identify two features of intellectual capital as being particularly important: attitude capital—"a predisposition to exploring cross-curricular links and formulating a constructively critical view of the benefits that they might bring"—and intellectual agility—"the ability to innovate and change practice, to think outside the box about problems and come up with novel solutions" (p. 56).

Clearly, a great deal of care needs to go into the planning of teacher education programmes, whether pre-service or in-service, if technology teachers are to be supported to collaborate more closely with their counterparts in other disciplines. For instance, Barlex et al. (2012) offer an example of a one-day workshop used to engage pre-service teachers of science and D&T in the UK in cross-curricula collaboration. A positive benefit for the students was the opportunity to experience and be part of 'collaborative creativity' such as that described by Lewis et al. (2007) above. However, more research is needed to analyse other examples of how pre-service and in-service technology teachers can be supported to work closely with other subject areas, and to evaluate the impacts on student learning.

## Imagining a Future with Greater Cross-Disciplinary Interactions

Given the traditional packaging of school learning into discrete subjects, with the concomitant organisational structures (timetabling, teacher education, high-stakes assessment processes), is it possible to imagine a future where technology is more closely aligned with other subjects? What might be the advantages of such alignment? What are the implications? Do current curriculum documents and assessment approaches support or discourage more cross-disciplinary approaches?

Mark Sanders (2012), in a keynote address to the Technology Education Research Conference in the Gold Coast, Australia, made a case for 'integrative STEM education' (iSTEM). As an outspoken critic of the overuse of 'STEM education' within political rhetoric and its lack of an agreed meaning (e.g., Sanders 2008), he proposes an alternative vision:

> Integrative STEM education refers to technological/engineering design-based learning approaches that intentionally integrate the concepts and practices of science and/or mathematics education with the concepts and practices of technology and engineering education. Integrative STEM education may be enhanced through further integration with other school subjects, such as language arts, social studies, art, etc. (Sanders and Wells 2006)

This vision—called 'iSTEM pedagogy'—"purposefully seeks to engage students in using/applying math, science, and engineering concepts and practices in designing, making, and evaluating solutions to authentic problems" (Sanders 2012, p. 110). The key, therefore, is that it is the technological context that drives the teaching and learning programme, rather than the technology being an 'add on' to a science or mathematics context.

In order for such a vision to be implemented, a paradigm shift is needed regarding how education is operationalised in schools. This shift will likely need to be bigger for secondary level than primary level, although even at the primary level—where one teacher tends to be responsible for students' learning across the curriculum—there are significant issues regarding teacher knowledge and the prioritisation of some subject areas over others. For example, New Zealand recently introduced the requirement for all primary schools to report on their students' achievement against national numeracy and literacy standards. This has tended to result in a greater valuing of numeracy and literacy learning over learning in other subject areas. In addition, funding for teacher professional development has been channelled away from other curriculum areas to focus on numeracy and literacy. While there is some exploration of how numeracy and literacy might be developed through other subjects, such as science or technology, this tends to occur only in isolated pockets. For technology, which is still considered to be a relatively new subject area despite having had a national curriculum since 1995, the decrease in targeted technology-specific professional learning funding is likely to have significant negative consequences. How can primary teachers be expected to align students' technology learning with their learning in other subject areas if they are unclear about what technology learning might actually look like?

At the secondary level, issues such as school timetabling, teacher collegiality and collaboration, and assessment all need to be considered if students' learning in technology is to become more aligned with their other learning, as indicated by the study by Lewis et al. (2007) introduced above. School structures may also need to be re-configured. How is the school day/week arranged? It is unlikely, for example, that significant project work can be seamlessly undertaken in isolated pockets of approximately one or two hours every few days. How do teachers collaborate in programme planning? Who is responsible for facilitating which parts of students' learning, and how are these decisions made? What professional learning support is required by teachers in terms of alternative pedagogical approaches? What opportunities are there for students to have input into what they will learn, how they will learn it, and how they will demonstrate their learning? Are there national or school-based assessment tasks that can be appropriated for assessment purposes, or will they impinge on the creativity of programme design? Each of these questions is likely to require significant change at the school and wider system levels.

## *'Impact Projects' at a New Senior Secondary School*

To demonstrate the possibilities for cross-curricular programme design, an example is offered of Albany Senior High School (ASHS), which opened in New Zealand in 2009 with the purpose of offering an innovative curriculum suited to preparing students for life and work in the twenty-first century. The intention of this example is to show what might be possible if we shift some of the traditional views of how schooling is organised.

ASHS is a purpose-built school in a prosperous, expanding urban environment. It caters for the three senior years of high school (Years 11–13), the feeder school (Years 7–10) also having recently been built. The intent, from the outset, was to offer a curriculum for the twenty-first century. This meant that learning would be engaging, accessible and useful for students, that teachers would be tutors, and that a disposition for lifelong learning would be fostered (Hipkins 2011). Learning spaces are open plan with easy access to computers as well as more traditional teaching tools such as whiteboards. Mixed teams of subject teachers work in each space. They also occupy the same workrooms. The arrangement is intended to maximise opportunities to build links across learning areas in order to enhance the coherence of the curriculum that students experience. This cross-curricular alignment is supported in the design of high-level course planning procedures and documents.

For four days each week, learning time is structured largely within traditional subject structures, although two slots are dedicated tutorial time when students meet in small groups with their designated tutor teacher. This tutor acts as a mentor of the allocated students' general academic progress, as well as the mentor for students' 'impact projects'. These projects are a key feature of the school's curriculum design. They are the focus of learning for one day each week, when traditional timetabling is suspended. The impact projects can be undertaken individually or in groups, are chosen by the students, and are designed to extend students' learning beyond what would be offered in class. The intention is that they provide students with challenging and personally meaningful opportunities for taking responsibility for their own learning. Parents or mentors from the school's wider community may be involved in supporting the students.

While obviously not all projects are technology-related, there is enormous scope for technology learning within the design of projects, and—with careful planning—for learning from other subject areas to be integrated and extended. The process undertaken by students, no matter what their specific impact project might be, is also aligned with important aspects of technological endeavours, including project design and planning, performing the plan, and evaluating the plan against success criteria. While a somewhat formulaic representation of technological endeavour, there is potential within this structure for student learning about the nature of technology.

New Zealand's senior secondary qualifications framework, the National Certificate in Educational Achievement (NCEA), is a standards-based system with a suite of achievement standards that can be flexibly combined. The vision is for schools to offer innovative programmes based on students' interests and needs. At ASHS, all teachers are familiar with other curriculum areas, including the technology curriculum and its related achievement standards, so that they can create an assessment package that is relevant and personalised to each student.

While it is less clear how the opportunities provided by NCEA for innovative assessment of impact projects have been operationalised, there is significant potential for supporting students' technological learning, including enhancing their technological literacy. However, this challenges traditional understandings of school subjects and requires careful negotiation with teachers, students, parents, and the wider community. Tempering innovation in course design and assessment is—unfortunately—University entrance requirements, which continue to operate as a conservative influence in that their systems are not yet designed to adequately accommodate the flexibility that is currently available within NCEA.

It needs to be said, too, that implementing these new curriculum innovations at ASHS has not been without its challenges (Hipkins 2011). Students were accustomed to teacher-directed learning, and some had difficulty choosing a topic and carrying out a relevant investigation. Students' relationships with their peers were tested, particularly when working together with friends. Teachers described themselves as being on a steep learning curve, learning to point students in the right direction without being too directive. The community needed to be persuaded of the value of the initiative. Provision of sufficient mentors requires ongoing creativity. Teachers and school leaders have needed to be courageous and resilient.

Fine-grained analysis of the students' learning—including aspects related to technological knowledge and skills—is still needed to understand how the curriculum structure is impacting on what is being learned and how that learning occurs. ASHS is also a new school. Documenting similar changes in existing schools, where pre-existing school cultures have a significant influence on ways in which a typical school day plays out, is needed to support school leaders in thinking about how they might continue to address the twenty-first century learning needs of their students and community.

## Concluding Thoughts

This chapter has adopted the view that a paradigm shift is needed in education to re-vision what 'schooling' would look like if students are to genuinely be supported to develop the knowledge, skills and dispositions they will need to effectively contribute in twenty-first century society and workplaces. One change that is likely to be necessary relates to a re-thinking of knowledge domains and the ways these have been traditionally separated into distinct school subjects. For example, 'wicked problems'—complex problems like climate change, natural hazards, public healthcare—are characterised by their multiple causes, their interdependencies, and the fact that they are difficult or even impossible to solve using the tools and techniques of one organisation or discipline.

If traditional subject siloes are to be re-conceptualised, one possibility is to situate students' learning within authentic, problem-based contexts that have personal meaning to them. Already, some schools have begun implementing such programmes, such as in the example presented of Albany Senior High School. Traditional timetable structures and school-community boundaries are suspended. Teachers take on the role of mentor or advisor. The learners play a large part in determining what they learn, how they learn it, and how they demonstrate their learning.

Embedding students' learning within a technology-based problem or task offers rich opportunities for students to develop technological skills and knowledge, as well as skills and knowledge in other discipline areas—science, for example, or social science or the arts. Opportunities for such diverse learning also reflects the contemporary nature of technological research and development, which is often carried out by a transdisciplinary collage of scientists, technologists, engineers and funding agencies. In addition, the interdependence between science, technology, the environment and the social and cultural context seems not only worthy of exploration within compulsory education, but necessary if students are to develop into technologically literate citizens, as advocated by contemporary school curricula. As Steve Keirl argued in Chap. 2:

> Whatever the framing, the delivery of any comprehensive technological literacy is unlikely to be possible through a single subject or learning area but, rather, would become the business of every teacher and department in a school.

Of course, teacher change is not easy, in part because of the dominating influence of subject subcultures. These often lead to a consensual view about the nature of the subject, the way it should be taught, the role of the teacher, and what is expected of students. Thus, a science teacher seeking to embed technological learning outcomes in a school programme is likely to be influenced by his/her background in science. In addition, there are indications that even when science teachers develop broader views of technology they may revert to their previously-held notions (e.g., of technology as applied science) when faced with disparities between their new views and their practice, or when entering areas of uncertainty (Jones 2012).

Given the traditional separation of science and technology as distinct school disciplines, supporting both pre-service and in-service teachers to consider the nature of both science and technology will require deliberate intervention. Finding teacher educators who themselves have a robust understanding of the similarities and differences between science and technology—and what this means for technology education—remains a significant challenge.

If teacher change is difficult, whole-school change is even more challenging to achieve. Albany Senior High School, as a new school, was in the fortunate position of being able to create an alternative model of 'doing school' from the outset. This impacted on the teachers they employed and the students that they attracted. It is important that the impacts of this alternative model on students' specific learning outcomes are identified, in part to persuade others of the value of thinking differently about how schooling might be done.

This book has deliberately set out to consider future issues and opportunities for technology education. One of the urgent considerations, explored in this chapter, is how technology education might be aligned with other subject areas—where 'alignment' is taken to mean the 'joining with others in a cause'. The cause, of course, is offering students a relevant, authentic, meaningful education, one that will equip them to contribute to a world where many of the challenges and opportunities will be associated with technological solutions.

In exploring implications of aligning technology with other subjects, aspects for teachers, school leaders and curriculum developers to consider are:

- the purpose of technology education at different levels of schooling (see Chap. 8, this volume, by David Barlex);
- teacher knowledge of the nature of technology;
- students' understandings of the nature of technology and whether it is primarily, for example, craftsmanship, innovation, ICT, the outcome from scientific endeavours, or a broad field of research and development in its own right;
- how technological learning outcomes might be authentically assessed (see Chap. 7 by Kay Stables);
- the focus for students' conceptual and procedural learning and the need to explicitly include technological knowledge and skills in planning and assessment if these are to be valued by the students and the teacher/mentor;
- students' identity formation and their views of themselves as 'technologists' and what they understand this to mean (see Chap. 3 by Marilyn Fleer);
- pedagogies to develop students' critical thinking and problem solving skills, as well as their innovation potential (see Chap. 5 by David Mioduser);
- the role of professional associations in promoting and maintaining the integrity of technology education to develop students' technological skills and understandings (see Chap. 13 by Kendall Starkweather); and
- the benefits and challenges of curriculum 'borrowing' and what messages might be adopted when looking at how others have sought to develop educational experiences that adequately prepare students for citizenship in the twenty-first century.

None of these aspects, on its own, is simplistic. When taken together, they represent a complex system of interacting elements. If formal education is to continue to shift in its beliefs and practices, there are rich opportunities for education researchers to work with teachers and school leaders to design and implement new learning programmes and analyse the impacts on students' experiences and learning outcomes as part of ongoing school reform.

## References


Barlex, D., Davies, S., & Hardy, A. (2012). Engaging trainee teachers of science and design & technology in cross curricula collaboration—A case study. In H. Middleton (Ed.), *Explorations of best practice in Technology, Design and Engineering Education* (Vol. I, pp. 102–117). Brisbane: Griffith Institute for Educational Research.

Broadfoot, P. (2000). Comparative education for the 21st century: Retrospect and prospect. *Comparative Education, 36*(3), 357–371.

Brown, J. S. (2006). New learning environments for the 21st century: Exploring the edge. *Change, 38*(5), 18–24.

Bull, A., & Gilbert, J. (2012). *Swimming out of our depth: Leading learning in 21st century schools.* Wellington: New Zealand Council for Educational Research.

Buntting, C., & Jones, A. (2009). Unpacking the interface between science, technology and the environment: Biotechnology as an example. In A. Jones & M. de Vries (Eds.), *Handbook of research and development in technology education* (pp. 275–285). Rotterdam/Boston/Taipei: Sense.

Campbell, C., & Jobling, W. (2012). Integrating technology and science—An opportunity missed. In H. Middleton (Ed.), *Explorations of best practice in Technology, Design and Engineering Education* (Vol. II, pp. 102–117). Brisbane: Griffith Institute for Educational Research.

Cowie, B., Moreland, J., & Otrel-Cass, K. (2013). *Expanding notions of assessment for learning. Inside science and technology primary classrooms.* Rotterdam/Boston/Taipei: Sense.

de Vries, M. (2009). The developing field of technology education: An introduction. In A. T. Jones & M. J. de Vries (Eds.), *International handbook of research and development in technology education* (pp. 1–9). Rotterdam/Boston/Taipei: Sense.

Gallagher, J. J. (1971). A broader base for science education. *Science Education, 55*(3), 329–338.

Gilbert, J. (2005). *Catching the knowledge wave? The knowledge society and the future of education.* Wellington: New Zealand Council for Educational Research.

Hipkins, R. (2011). *Learning to be a new school: Building a curriculum for new times.* Wellington: New Zealand Council for Educational Research.

Jones, A. (2012). Technology in science education: Context, contestation, and connection. In B. J. Fraser, K. Tobin, & C. J. McRobbie (Eds.), *Second international handbook of science education* (pp. 811–821). Dordrecht: Springer.

Jones, A., & Buntting, C. (2013). The role and place of science and technology education in developing innovation education. In L. V. Shavinina (Ed.), *Handbook on innovation education* (pp. 419–429). New York: Routledge.

Jones, A., Buntting, C., & de Vries, M. (2013). The developing field of technology education: A review to look forward. *International Journal of Technology and Design Education, 23*, 191–212.

Kelly, G., Luke, A., & Green, J. (2008). What counts as knowledge in educational settings: Disciplinary knowledge, assessment and curriculum. *Review of Research in Education, 32*, vii–x.

Lewis, T., Barlex, D., & Chapman, C. (2007). Investigating interaction between science and design & technology (D&T) in the secondary school—A case study approach. *Research in Science and Technological Education, 25*(1), 37–58.

Roth, W.-M., Tobin, K., & Ritchie, S. (2001). *Re/constructing elementary science.* New York: Peter Lang Publishing.

Sanders, M. E. (2008). STEM, STEM education, STEMmania. *The Technology Teacher, 68*(4), 20–26.

Sanders, M. (2012). Integrative STEM education as 'best practice'. In H. Middleton (Ed.), *Explorations of best practice in Technology, Design and Engineering Education* (Vol. II, pp. 102–117). Brisbane: Griffith Institute for Educational Research.

Sanders, M. E., & Wells, J. G. (2006). *Integrative STEM education.* Retrieved January 8, 2013, from http://www.soe.vt.edu/istemed/

Shavinina, L. V. (Ed.). (2013). *Handbook on innovation education.* New York: Routledge.
Williams, P. J. (2002). Processes of science and technology: A rationale for cooperation or separation. In I. Mottier & M. J. de Vries (Eds.), *Proceedings of PATT12. Technology education in the curriculum: Relationships with other subjects* (pp. 33–50). Accessed June 12, 2013, from http://www.iteea.org/Conference/PATT/PATT12/PATT12.pdf
Young, M. (2010). Alternative educational futures for a knowledge society. *European Educational Research Journal, 9*(1), 1–13.

# Chapter 11
# Vocational and General Technology Education

**P John Williams**

This chapter explores the interactions between vocational and general approaches to Technology Education, proposing that the vocational–general divide is not always clear. Often, this reflects the tension between training for specific skills competencies, and educating for more generic core competencies. With schools increasingly being held accountable for the performance of their graduates, and rising unemployment, Technology Education as strictly general education is being questioned and vocational education is being infused with liberal arts characteristics. Globalisation, rapid changes in technologies and work places, and growing understanding of the nature of learning and transfer of learning will all influence the future Technology Education curriculum. While integration of vocational and general approaches is problematic at the level of single classes, at the programme level it may provide meaningful education pathways.

## Introduction

This chapter is concerned with issues arising from the vocational and general elements of Technology Education. In many countries Technology Education is a component of the core curriculum in both primary and secondary schools, providing a sequence of experiences that are judged to be useful for all students in preparing them to play a full role in society and to achieve to their full capability. In addition, many schools offer students more specific forms of Technology Education focussed on a particular vocation or group of vocations. These classes are not intended for all students, but only those interested in pursuing a specific vocation. In developed

PJ. Williams (✉)
Technology, Environmental, Mathematics and Science Education Research Centre, University of Waikato, Hamilton, New Zealand
e-mail: pj.williams@waikato.ac.nz



PJ. Williams et al. (eds.), *The Future of Technology Education*, Contemporary Issues in Technology Education, DOI 10.1007/978-981-287-170-1_11

countries, these vocational options tend to be offered at the upper secondary level; in less developed countries they may be offered at lower levels of schooling.

However, the vocational–general divide is not always clear. Many subject areas that are now included in the technology learning area at the secondary level have had a vocational orientation in the past, such as those subjects dealing with hard materials. These subjects tended to have a quasi-vocational status: while they were seen as providing an orientation to various vocations, there was no explicit connection with industry, and no industrial accreditation.

In addition, in many ways there is a blurring of the boundaries between the two approaches to Technology Education. They are often taught in the same facilities to the same students, it is often the same technology teachers who teach both general and vocational subjects, and the transposition between the two approaches to Technology Education is not always easy.

The trend from a vocational to a general approach to Technology Education in schools, and the differentiation between the two, has taken place over a long period of time, and in many instances, still continues—often resulting in confusion and tensions. The popular inclusion of vocational technology options at post-compulsory levels of schooling—particularly when attempts have been made to offer the two approaches simultaneously—has necessitated a clear rationale for both approaches (see Chap. 8, this volume, by David Barlex).

Issues arising from the differentiation between vocational and general Technology Education are explored in this chapter. Also considered is the notion of transfer, which is fundamental to the effectiveness of vocational education. When the nature of learning is considered in the context of the transfer of capabilities from one context to another, a complex interaction of factors comes into play.

## Definitions

The first form of Technology Education, in early civilisation, was probably vocational, involving the development of competencies using technological artefacts to achieve specific goals. Competency testing was probably quite rigorous, and judged by whether one survived or not. The system of competency training would have been informal, although there may have been small groups of people who practiced with different tools under a master–apprentice type of structure. Technology Education as a form of general education came much later, and quite recently (maybe the past 30 years) was accompanied by the recognition that society is essentially technological, and in order for schools to prepare students for such a technological society, Technology Education needs to be included as part of the school curriculum. 'Technology' is the most recent iteration of this educational focus; prior to this it was 'Industry' that was recognised as a valid focus, and prior to that 'manual skills' were seen to be important for all students to develop.

The *study* of technology as a vocational or a general approach is longstanding. Maclean and Wilson (2009) consider that "the study of vocational education has a longstanding history, beginning in the 1880's when urbanisation, mechanisation and industrialisation became the major forces driving societies" (p. lxxxviii). However, it could probably be traced a lot further back, maybe to the ideas of Comenius and Locke in the 1600s, and then Pestalozzi and Froebel in the early 1800s, who all developed theoretical positions on aspects of what we now call Technology Education.

The approach taken in this chapter is that Technology Education is the broad curriculum area, under which a vocational or a general approach can be taken to the delivery of content. The level of vocational education that will be discussed in this chapter is that which occurs in schools. While this narrows the field of discussion somewhat, there is still significant diversity across international education systems. This diversity ranges from unstructured work preparation experiences offered to primary school students in developing countries for whom the completion of primary school represents the end of their formal education, to structured and externally-certified attainment of specific competencies at the upper secondary level, which are then transferred to advanced standing credit toward a tertiary qualification.

Despite the agreement at the second International Congress on Technical and Vocational Education in 1999 that the term Technical and Vocational Education and Training (TVET) should be used as a way to unite the field (UNESCO 1999), there remains a diversity of terminology. Some terms are used in particular geographical areas, while others represent specific characteristics. The terms include technical/vocational education (TVE), vocational education and training (VET), career and technical education (CTE), occupational education (OE) and continuing vocational education and training (CVET). The definition of TVET adopted by UNESCO (1999) is "the study of technologies and related sciences, and the acquisition of practical skills, attitudes, understanding and knowledge relating to occupants in various sectors of economic and social life" (p. 2).

A similar level of diversity exists in the provision of Technology Education as a component of general education. Many jurisdictions around the world have a K–12 (early years to the end of secondary) structure for the delivery of Technology Education. However, others combine technology education with science at some levels, separate primary and secondary structures, or only address some levels within the K–12 spectrum. Adding to the diversity, what actually happens in schools may bear little resemblance to formal curriculum requirements, with some schools paying lip-service to the curriculum and others making technology a school priority and using it as an integrating mechanism across the curriculum. The meaning adopted in this chapter is that Technology Education as general education is the study of technology in which students learn about the processes and knowledge of technology in order to develop their technological literacy. This is accomplished through exploration and experience of a wide range of technologies in a variety of contexts in which students work creatively and analytically to critique, design and develop products and systems.

## Response to Context

The contexts to which both vocational and general education respond are dynamic. Vocational Technology Education responds to the needs of a range of industries, which in a global sense change slowly over time but in a national or regional sense can change more rapidly as economic priorities change. General Technology Education is responsive to the technological nature of society, which changes more rapidly as the technology innovation–infusion cycle becomes increasingly shorter. An issue for both these approaches to Technology Education is that educational structures (curriculum, equipment, facilities) change slowly. There is not the economic imperative for rapid change in education that drives commercial sectors. From this differentiation, a number of issues arise.

All countries are at different stages of social and economic development, and the focus on vocational education reflects the stage of development. Those developing countries striving for a basic level of literacy for all may not have a focus on vocational education, or may integrate it into primary education (see Chap. 12, this volume, by Frank Banks and Vanwyk Chikasanda, for a discussion on technology education in Bangladesh and Malawi). On the other hand, as countries progress through universal access to secondary education, there may be a focus on the provision of post-secondary pathways in vocational areas. There is some evidence that this focus (i.e., low income countries focusing on primary education) represents the best returns on investment for country development (ADB 2009).

Another context that increasingly demands an educational response is related to globalisation. As technology facilitates the movement of jobs from countries with an excess of skilled labour (USA and some European countries) to those with an excess of low skilled labour (India and China), vocational education plays a role in both. It is a means of increasing the skill level of the population in the former countries and so maintaining acceptably low levels of unemployment; in the latter countries it is a fundamental plank in the economic development of the nation.

Statistics seem to indicate a positive correlation between vocational education as a national component of human resource development, and national economic growth (Sabadie and Johansen 2010). However, there are some significant exceptions. On one end of the spectrum is Japan, with high GDP per capita (US$28,000) but low participation of secondary students in vocational education (25 %); at the other end there is Indonesia, with high participation rates (37 %) but low GDP per capita (US$2,600).

An alternative view was put forward by Pritchett (2006) after analysing rich and poor countries over 27 years: that economic growth precedes education, rather than the reverse. Chang (2010) also concluded that there was very little evidence that more education (resulting in higher rates of literacy) leads to greater prosperity, after analysing the relationship between literacy rates and per capita income. Another alternative view is that information learned in school has little impact on worker productivity, even in jobs where the application of a degree is obvious—a mathematics degree in investment banking, for example:

> Employers hire university graduates over high school graduates because a college degree suggests general intelligence, self-discipline, and organization. It's not what you've learned, just the fact that you went to college, got passing grades and graduated that counts—specialized knowledge is usually irrelevant. (McNerney 2013)

Nevertheless, the argument for the provision of vocational education as a form of human capital development is a persuasive one, and as a mechanism for both developing countries and countries requiring significant economic adjustment to stimulate economic growth, the provision of vocational education is a response to these needs.

## Specific and General Competencies

Within vocational education, there seems to be a tension between training for specific skills competencies and more generic core competencies. On the one hand, many industry groups and individual employers state that they want general competencies in new employees, such as working with others and in teams, solving problems, communicating ideas and information, and using technology (Mayer 1992). Many countries have consolidated lists of key general competencies for employment—variously called workplace knowhow, life skills, essential skills or employability skills—which tend to encompass similar types of attributes and generally address communication, problem solving, self management, numeracy, information technology, and work ethics. In 2008, the establishment of the Asia-Europe Meeting research network (http://www.aseminfoboard.org/) to identify core competencies and to explore the ways in which they operate is an indicator of a developing focus on general competencies. There is also some evidence that students tend to value their general employability skills, rather than their specific competencies after high school completion (Bowskill 2012).

On the other hand, individual industries develop lists of specific competencies that are required to be mastered in order to obtain a qualification in that industry. For example, in the Australian Metals and Engineering training package, competency MEM8.3C includes:

Elements and Performance Criteria

1. Identify electroplating requirements
   - 1.1 Electroplating requirements are identified.
   - 1.2 Untreated materials and required electroplating treatment are identified.
2. Prepare for electroplating process
   - 2.1 Materials and racking arrangement are checked for non-conformance to specifications/job requirements.
   - 2.2 All plant and equipment relevant to process are checked for compliance with safety and operational requirements.
   - 2.3 Instrumentation/gauges are checked for operation.
   - 2.4 Condition of solution is checked.
3. Perform electroplating
   - 3.1 Operation steps are carried out in correct sequence

The assessment structure privileges the specific competencies. As mastery of these competencies needs to be verified in order to grant the qualification, this is the focus of both the teaching and assessment. General competencies are not specifically assessed, despite employers stating that these are the most important, and many industries support specific skills training once they employ someone.

Some attempts to foreground the general competencies are occurring. For example, in Singapore, specific courses are offered in the ten foundation areas (literacy and numeracy, communication, problem solving, initiative and enterprise, communications, lifelong learning, global mindset, self management, life skills, and safety) as a form of alternative entry into technology certificate courses for those without formal qualifications, but this approach is not common.

In summary, while vocational assessment and certification are based on specific competencies, employers are more interested in general competencies—although these are not often explicitly addressed by schools.

## Nature of Learning and Transfer

Research on the transfer of learning indicates that transfer is difficult to achieve for many reasons (Perkins and Salomon 1992). While it is obvious many skills, such as literacy and numeracy, can be taught for transferability, the mastery of other skills in a school context is no guarantee that a specific skill can be transferred to a different school context or to a situation outside of school. This is problematic for vocational *and* general technology education, where the foundational validity for the subject's existence rests on notions of transferability—either the transfer of specific skills and competencies to a workplace context, or the exercise of cognitive skills, such as problem solving, critiquing and thinking creatively in a range of situations outside of school.

Traditional notions of transfer were based on the work of Thorndike (1924) and argued that if the situation to which the skills or knowledge is being applied is similar to that in which it was learnt, then transfer would be automatic. Research has since indicated that this understanding is overly simplistic and mechanistic, and does not consider factors such as the cognition that is needed to support transfer, the context of the learning environment, the generalisability of learning principles, and the nature of learning.

An elaboration of Thorndike's (1924) notion of transfer is that of near and far transfer (Perkins and Salomon 1996) in which near transfer relates to the contextual similarity of the learning and the applied situations. Conversely, far transfer is the application of learning into a context which is quite dissimilar from that in which the knowledge was gained, and the dissimilarity implies that far transfer is more difficult that near transfer. Leberman et al. (2006) point out that far transfer is becoming more critical because of the rapid changes in knowledge, technology and workforce opportunities, rendering many workplaces increasingly different from school contexts.

In the early 1990s there was recognition that the development of skill practice with a view to transfer required more than just the repetition of tasks until mastery was achieved (Griffiths and Guile 2004). Inquiry-based learning was consequently introduced to support learners to critically observe their work and reflect on their observations. Allied with Kolb's (1984) idea of the experiential learning cycle, this led to broader understandings of the nature of learning for transfer, which related to the individual's personal and social development. This is supported by Lave and Wenger (1991) who assert that situated learning is neither an educational form nor a pedagogical strategy, but rather legitimate participation in a community of practice. In other words active engagement in, and acceptance into a community ensures transfer into that community.

Johnson et al. (2011) contend that any notion of transfer that does not consider cognitive ability and function, but just focuses on the context, is inadequate. They elaborate on metacognition, mental representations and analogical reasoning as important cognitive concepts which contribute to successful transfer, and that teaching for transfer involves cognitive attention at both the initial and the receptive ends of the transfer spectrum: "Teaching for transfer involves linking new knowledge to existing schemata, naïve theories and mental models of students, and reorganizing these cognitive structures where necessary" (p. 64).

Griffiths and Guile (2004) drew on literature from sociology, management and innovation systems in an attempt to theorise the new economic and technological conditions for shaping the knowledge economy and how this impacts on vocational learning for transfer. One of their conclusions was that learners need to learn how to draw on their theoretical knowledge to interrogate workplace practices, and on their everyday knowledge to interrogate theory. Without this iterative process of interrogation, or metacognition, vocational learning will not easily transfer.

An alternative framework (though sympathetic to these previous notions) through which to analyse transfer is that of activity theory (Engeström 1987): that learning is not really being transferred from one context to another, but the individual is continually learning through one changing situation to the next, or from one activity system to the next, each one being increasingly complex. From this view, the ability to move between different activity systems (e.g., school and workplace) and become active and useful members of each system reflects the ability to transfer knowledge between contexts. While the basis of the analysis of transfer is the system rather than the individual, the unit of analysis is the activity itself, which takes into consideration social and cultural aspects of the setting.

This leads to Beach's (2003) notion of consequential transition, in which the application of knowledge is never decontextualised from social organisation. There are parallels here with Vygotsky's notions of social constructivism. The elements of transfer become the sequence of processes that form the interplay between individuals and social organisations, both of which are dynamic and can be represented by the fluidity of an activity system.

Consequently, Guile and Young (2003) argue that in preparing future employees for the workplaces of contemporary and global economies, the focus should be on distributed work rather than specific and narrow skills and competencies.

Student participation in a range of activity systems (e.g., school and workplace) helps develop boundary-crossing skills through the modification of, and contribution to, daily practices that are a way of developing knowledge. Students therefore should be actively involved in their learning systems, rather than simply copying skills or learning information. This provides the rationale for school programmes such as day release for students into industry, which gives them the opportunity to experience a range of social organisations. However, without scaffolding, a student's ability to integrate and become an active and effective member of each system is not guaranteed.

This logic leads to an observation by Säljö (2003) that there may be no need for theories related to transfer because such notions are accommodated within theories of learning. Take, for example, Piaget's concepts of cognitive constructivism and Vygotsky's social constructivism in which learners actively construct knowledge through assimilation and accommodation in the former (Bettencourt 1993), and collaboration and social interaction in the latter (Powell and Kalina 2009).

In developing their own skills and understandings, students are cognitively active in applying critical thinking processes to solving problems or responding to issues through the application of skills. The information they draw on originates from many sources—their teachers, information technologies, experts, social others, and their own previous experiences. The information which is applied from their own experiences or understandings is, in effect, transfer. This notion of transfer, as an integral aspect of learning, is only valid in contexts of active learning. If the learning is passive, or the learner is not fully engaged, then neither the skills copied off the instructor nor the rote learning of information will facilitate ease of transfer. Thus, the learner must be critically engaged in both the development of learning to provide the basis for facilitating transfer, and also in the learning to which prior understandings and skills are applied. If either context is not an activity (in terms of an activity system) then transfer is problematic, and in this context, alternative notions of transfer need to be developed. However, if students are active in their learning, then constructing knowledge across the boundaries of activity systems encompasses notions of transfer.

Within a concept of social constructivism, it is not specific skills or knowledge that are transferred, but the reconstructed incarnation of the original skills or knowledge that is critically applied to the new learning context. The implication is that learning must be active, designerly, inquiry-based, and problematic (or some other form which fosters engagement) in order for students to have the mindset and ability of reconstruction.

A framework aligned to that of social constructivism is Wenger's (1998) communities of practice, in which social participation in a community is essential for learning to take place. Through interaction with members of a community, an individual comes to learn the practices of that community and develops an identity as a member. The concepts of 'practice' and 'participation' are critical to this framework in conveying the notion of *active* engagement in the community. The framework does not support passive learning or rote repetition of skills and understandings, with the implication that this does not represent true learning and, in the context of this discussion, would not facilitate transfer.

Transfer, from a community of practice perspective, involves the individual moving from one community to the next, and integration into that community involves the adoption of the rules and practices of the new community. Vocational and Technology Education school classes are communities of practice in which the possession of skills, rules and understandings are characteristics of membership of the community. One of the rationales for the existence of these communities is to ease the transfer of individual members into other related communities. For example, a VTE school community that focuses on the development of skills and understandings related to hospitality is preparing its members for transition into communities of practice such as hotel management, tourism planning or bartending. The similarities of the communities being transitioned, and the shared concern and passion for what is done within a community (Krishnaveni and Sujatha 2012) facilitate the application and restructure of prior learning.

The need for 'work process knowledge' has become a feature of some EU vocational systems (Boreham and Fischer 2002), in which product data, the organisation of labour, the social ecology and systems of the workplace have been incorporated into educational programmes. Such an approach recognises that contextual understanding is an important facilitator of transfer, together with a combination of theoretical and practical learning.

While the notion of transfer is explored in different ways within various sociocultural frameworks, as described above, there are some commonalities (Crafter and Maunder 2012):

- Transfer is a complex and fluid process of individuals and contexts involving the reapplication or restructuring of learning rather than its direct 'copy and paste' transfer.
- Active social interactions are a crucial factor in the initial learning stage and also in the applied context.
- Transfer is individually transformative and the processes of reflection and identity construction change the individual as they cross boundaries and move between communities.

A theoretical model of school to work transition based on social cognitive career theory (Lent et al. 1999) was used by Masdonati (2010) to research student readiness and success in transiting to employment. The model comprised elements of self perception, perception of social support, and institutional resources and barriers. A programme based on the model improved participant readiness for the workplace.

Instead, then, of viewing transfer as the re-application of skills and knowledge from one context to another, it should be seen as a process of boundary crossing that involves consequential transitions in which learners are engaged in a variety of quite different social, intellectual and manipulative tasks in different contexts. Effective transfer, then, involves theory and practice of the skills and knowledge of the vocational domain, self organisation and workplace enculturation, and mediation between these relationships.

## Role of ICT

Information and communication technologies (ICTs) are playing an increasingly significant role in education, and the indicators are that this will continue to include vocational education. While the content of many vocational areas becomes increasingly complex, the synergies in presenting technology through the use of (information) technology become clear. Having now bypassed the early and ill-informed arguments about the cost of online provision of learning resources, the need for appropriate pedagogies and the role of the teacher, the use of ICT can be informed and targeted.

Information technologies are enabling the elimination of boundaries that were shaped by the forces of previous paradigms. For example, the categories of 'developing' and 'developed' countries are less of a distinction in the global knowledge-based economy in which global networks are becoming more of an organising principle. The move from a knowledge society to a social society is reflected in the increasing use of social media and networking.

ICT literacy is a necessary component of both vocational and general technology education. It constitutes one of the general competencies of vocational education, in preparation for an increasingly digital and competitive workplace market. It is also an aspect of technological literacy for all students, to enable continual improvement and lifelong learning.

Further, as the rate of technological change continues to increase and globalisation pressures come to bear on many commercial sectors, changes in workforce requirements become more dynamic, and consequently the workforce needs to be flexible. Career changers can generally ill-afford time off from employment to re-train, so the affordances of ICT in delivering information to enable workforce transitions are relevant.

The reasons for incorporating ICT into the delivery of technology courses are similar to the more general reasons for incorporating digital technologies into learning—flexibility, experiential opportunities, cost savings, and the integration of theory and practice (Manir 2009). Studies show that e-learning is being used to complement and support traditional learning with the outcome of a blended form of learning rather than fully online courses.

There are advantages to incorporating e-learning into technology education because of the nature of the subject area. It is possible to develop sophisticated simulations of situations that students would not otherwise be able to access. For example, there is a safety advantage in developing a familiarity with dangerous tools and machinery, and while simulations will not enable the developments of competencies they do provide a level of awareness that may not otherwise be possible.

A perennial issue for technology education, particularly vocational, is the cost of developing in students competencies that require access to sophisticated and expensive equipment and machinery. Of course business and industry have a commercial rationale for the purchase of equipment, which is not generally available to educational institutions. e-Education may provide a partial solution to this problem through the provision of virtual experiences to students.

## Convergence of the Vocational and General

Even when focussing on the development of technical skills, a relevant and vigorous vocational education community will engender a range of other attributes. For example, after meeting students from a successful agricultural science programme in Arizona, Klein (2012) observed that "These students also knew how to make an impression; they had learned the soft skills necessary to be good employees. They looked you in the eye, introduced themselves and shook your hand."

I made a case in 1998 (Williams 1998) that in Technology Education, the generalisation of vocational education and the vocationalisation of general education was resulting in the confluence of their goals. I maintain that this confluence continues. The requirements of progressive industries for new employees are similar to the desired outcomes of Technology Education as a component of general education. There could be a number of reasons for this harmony in development, in that similar factors have impacted on both education and industry. For example:

- increasing significance of technology in society,
- common economic conditions providing the stimulus for change,
- globalisation,
- increasing unemployment and increasing student retention rates,
- the recognition of a diverse range of learning styles,
- increasing quality of teaching and learning,
- the need for accountability, and
- social pressures on schools for graduate employability.

These influences have resulted in the entry-level core competencies of industry and the nature of general Technology Education having a number of parallels, such as:

- multidisciplinary in nature,
- standards developed as a guide to quality,
- similar personal development goals, such as flexibility, creativity, critical thinking, innovative and adaptable,
- individual responsibility for personal development, and
- team work skills.

In other words, there are two broad initiatives that to a certain extent overlap, but have been developed by different sectors of society: the changes and developments in Technology Education have been largely influenced by educators; the descriptions of the core competencies that are required at entry level in industry have been developed largely by industry. In both initiatives, politicians have played a significant role, but mainly in instigating the developments. The types of outcomes of these developments have been similar.

With schools increasingly being held accountable for the performance of their graduates, and unemployment rising, Eechnology Education as strictly general education is being questioned and vocational education is being infused with liberal arts characteristics. It may be necessary to equate general education more with vocational *education*, as distinct from vocational training. Discussions about these and similar terms reflect a movement to infuse technical education with some of the traditional liberal arts characteristics. Higgerson and Rehwaldt (1990) support this with the following types of arguments:

- technological changes are so rapid that any education that is limited to technical training is soon outdated; how to think and learn for future self education is more important,
- professionals need a combination of the liberal arts and the useful arts (Boyer 1987),
- technical education largely ignores the human factor but almost all workers need to deal with people, and
- technical training is inherently specialised, but technology is contextualised.

Others identify similar changes over time, but within a different frame. For example, Pavlova and Maclean (2013) attribute changes in vocational education to countries shifting focus from social concerns, whereby vocational programmes helped promote the inclusion of less privileged groups, narrowed education gaps and avoided social fragmentation; to a focus on economic issues within which vocational programmes become part of a human resource development agenda in order to enhance economic development. According to Karmel (2007) this has resulted in a VET system that has become industry-led rather than educationally driven.

Nevertheless, there seems to be a convergence of vocational and general technology education goals. This has implications for the ways schools structure their Technology Education offerings.

## Integration of Vocational and General

There are two levels of integration that relate vocational and general Technology Education approaches to each other. The first is classroom integration and involves teaching for both vocational and general outcomes in the same class. The second is at a programme level, where students are able to select from vocational and general options to construct a course that suits them.

### *Classroom Integration*

Teaching for both vocational and general outcomes in the same class is problematic, although for many schools it is perceived as a way of maximising the options available to students in cases where there may not otherwise be sufficient students to make up full classes.

The approach occurs where there is overlap between the vocational and general technologies that are offered, for example, in an engineering design course, which may be offered as a general education option, but which will cover some of the competencies needed for a vocational qualification, such as aspects of fabrication, for example. This connectedness is rarely a complete match in that the general education approach would be unlikely to cover all of the competencies required for a vocational qualification simply because the goals of the two approaches are different. The organisation of content would also be different. In a vocational approach, a structured set of activities would be organised so as to provide opportunities for the demonstration of mastery of competencies, whereas in the general, design-based approach, the content needed in order to address the design problem may not be known at the beginning of the class.

Of course, assessment varies between systems, but for a vocational approach that is part of a qualification, it is often in the form of mastery of competencies. Students are provided with a number of opportunities to demonstrate mastery, and when achieved, it is so recorded. Assessment for general education purposes would typically be different and could relate to achievement of an outcome or a standard at a particular level where judgements might be made based on a variety of sources of evidence, such as a portfolio, interview and project.

The pedagogical approach taken in a vocational class will also typically be different from that used in a general class. With a focus on vocational competencies, there is less opportunity for a student-centred approach because all students are working toward the same set of externally devised competencies. However, in a more design-oriented general approach, each student may be working on a different problem, or interpretation of a problem, so the prevailing pedagogy essentially becomes student centred.

In other words, the content (general vs. specific), goals (general vs. vocational), pedagogy (student centred vs. other centred), and assessment (outcomes vs. competencies) are quite different, and to try and combine them will result in doing neither well.

## *Programme Integration*

As a way to enhance the quality of delivery of vocational education, and to address the findings of Polesel (2008) that, at least in Australia, most vocational programmes are poor quality and do not provide students with either general or specific vocational competencies, Pavlova and Maclean (2013) suggest the inclusion of a component of general education "which focuses on the development of general/employability skills . . . as well as linking theory and practice to develop a holistic understanding of practices by the students" (p. 49). This second form of integration is more at a programme level, where students can take subjects from both vocational and general streams to devise a programme of study that suits them. For example, in the Republic of Korea, 40 % of secondary students are enrolled in vocational classes, and in some schools vocational and general education students share as much as 75 % of a common curriculum. Part of the rationale for this is to overcome the perception that studying a vocational sequence of subjects precludes later entry into other forms of tertiary education.

An increasing proportion of the workforce are in positions where there is no guarantee of lifetime job security. In this context, developing competencies for a specific vocation seems a bit short sighted, and the integration of general education into vocational programmes may be a more useful preparation for employment opportunities that unfold over time. As a result, particularly advanced countries are making upper secondary vocational programmes more general by integrating more academic content so that students can be more versatile in the occupational options they consider.

## Conclusion

Based on this discussion of issues surrounding vocational education, it is possible to make some reasonable projections into the future:

- Vocational programmes will become more student oriented as a broader range of subjects become available for selection across both vocational and general technology areas.
- The organisation of vocational education will become less centralised, which will enable schools to develop more flexible responses to local labour market needs.
- The provision of vocational education will become more flexible through the use of electronic networks. ICTs will enable a broader range of experiences, and simulations will provide a deeper awareness of vocational contexts.
- Deeper understandings of the nature of learning will facilitate transfer of attitudes, skills and competencies more effectively.
- The recognition that vocational education will need to become more generic and that proficiency in technical and para-professional skills will not be enough in most jobs will have the effect of standardising Technology Education to focus on generic skills.

## References

ADB. (2009). *Education and skills: Strategies for accelerated development in Asia and the Pacific*. Metro Manila: Asian Development Bank.

Beach, K. (2003). Consequential transitions: A developmental view of knowledge propagation through social organizations. In T. Tuomi-Grohn & Y. Engeström (Eds.), *Between school and work: New perspectives on transfer and boundary-crossing* (pp. 39–61). Boston: Pergamon.

Bettencourt, A. (1993). The construction of knowledge: A radical constructivist view. In K. Tobin (Ed.), *The practice of constructivism in science education* (pp. 39–50). Washington, DC: AAAS.

Boreham, N., & Fischer, M. (2002). *Work process knowledge*. London: Routledge.

Bowskill, N. (2012). *Five case studies exploring the value of technology education in New Zealand secondary schools*. Unpublished MEd thesis, The University of Waikato, New Zealand.

Boyer, E. L. (1987). *College: The undergraduate experience in America*. New York: Harper and Rowe.

Chang, H. (2010). *23 things they don't tell you about capitalism*. New York: Bloomsbury.

Crafter, S., & Maunder, R. (2012). Understanding transitions using a sociocultural framework. *Educational and Child Psychology, 29*(1), 10–18.

Engeström, Y. (1987). *Learning by expanding: An activity-theoretical approach to developmental research*. Helsinki: Orienta-Konsultit Oy.

Griffiths, T., & Guile, D. (2004). *Learning through work experience for the knowledge economy* (Cedefop Reference series 48). Luxembourg: Office for Official Publications of the European Communities.

Guile, D., & Young, M. (2003). Transfer and transition between education and work: Some theoretical questions and issues. In T. Tuomi-Grohn & Y. Engeström (Eds.), *Between school and work: New perspectives on transfer and boundary-crossing* (pp. 63–81). Boston: Pergamon.

Higgerson, H. G., & Rehwaldt, S. (1990). Interrelationships: Liberal education and technical training. *Journal of Studies in Technical Careers, 12*(3), 195–203.

Johnson, S., Dixon, R., Daugherty, J., & Lawanto, O. (2011). General versus specific intellectual competencies. In M. Barak & M. Hacker (Eds.), *Fostering human development through engineering and technology education* (pp. 55–71). Rotterdam: Sense.

Karmel, T. (2007). Vocational education and training in Australian schools. *The Australian Educational Researcher, 34*(3), 101–117.

Klein, J. (2012). Learning that works. *Time*, May 14. Accessed 1 Feb 2013 from: http://content.time.com/time/magazine/article/0,9171,2113794,00.html

Kolb, D. (1984). *Experiential learning*. Englewood Cliffs: Prentice Hall.

Krishnaveni, R., & Sujatha, R. (2012). Communities of practice: An influencing factor for effective knowledge transfer in organizations. *The IUP Journal of Knowledge Management, 10*(1), 26–40.

Lave, J., & Wenger, E. (1991). *Situated learning: Legitimate peripheral participation*. Cambridge: Cambridge University.

Leberman, S., McDonald, L., & Doyle, S. (2006). *The transfer of learning*. Burlington: Gower.

Lent, R., Hackett, G., & Brown, S. (1999). A social cognitive view of school to work transition. *The Career Development Quarterly, 47*, 312–325.

Maclean, R., & Wilson, D. (2009). Introduction. In R. Maclean & D. Wilson (Eds.), *International handbook of education for the changing world of work* (pp. lxxiii–cxii). Dordrecht: Springer.

Manir, A. K. (2009, June 2–5). *ICT competency framework for library and information science schools in Nigeria: the need for model curriculum*. Paper presented at the NALISE conference held at UNN.

Masdonati, J. (2010). The transition from school to vocational education and training. *Journal of Employment Counselling, 47*(1), 20–29.

Mayer, E. (1992). *Employment-related key competencies*. Canberra: Australian Education Council.

McNerney, S. (2013). *We don't need no education*. Accessed 1 Feb 2013 from: http://bigthink.com/insights-of-genius/we-dont-need-no-education

Pavlova, M., & Maclean, R. (2013). Vocationalisation of secondary and tertiary education: Challenges and possible future directions. In R. Maclean, S. Jagannathan, & J. Sarvi (Eds.), *Skills development for inclusive and sustainable growth in developing Asia-Pacific* (pp. 43–66). Dordrecht: Springer.

Perkins, D. N., & Salomon, G. (1992). Transfer of learning. In T. Husen & T. N. Postelwhite (Eds.), *International handbook of educational research* (2nd ed., Vol. 11, pp. 6452–6457). Oxford: Pergamon.

Perkins, D., & Salomon, G. (1996). Learning transfer. In A. C. Tuijnman (Ed.), *International encyclopedia of adult education and training* (pp. 422–427). Tarrytown: Pergamon Press.

Polesel, J. (2008). Democratising the curriculum or training the children of the poor: School-based vocational training in Australia. *Journal of Education Policy, 23*, 615–632.

Powell, K. C., & Kalina, C. J. (2009). Cognitive and social constructivism: Developing tools for an effective classroom. *Education, 130*(2), 241–249.

Pritchett, L. (2006). *Let their people come: Breaking the gridlock on global labor mobility.* Washington, DC: Centre for Global Development.

Sabadie, J. A., & Johansen, J. (2010). How do national economic competitiveness indices view human capital? *European Journal of Education, 45*(2), 236–258.

Säljö, R. (2003). From transfer to boundary crossing. In T. Tuomi-Grohn & Y. Engeström (Eds.), *Between school and work: New perspectives on transfer and boundary-crossing* (pp. 311–321). Boston: Pergamon.

Thorndike, E. (1924). Mental discipline in high school studies. *Journal of Educational Psychology, 15*, 1–22.

UNESCO. (1999). *Lifelong learning and training: A bridge to the future.* (Final report of the second international Congress on TVET, Seoul, 1999.) Paris: UNESCO.

Wenger, E. (1998). *Communities of practice: Learning, meaning, and identity.* Cambridge: Cambridge University.

Williams, P. J. (1998). The confluence of the goals of technology education and the needs of industry: An Australian case study with international application. *International Journal of Technology and Design Education, 8*, 1–13.

# Chapter 12
# Technology Education and Developing Countries

**Frank Banks and Vanwyk K.M. Chikasanda**

Bangladesh and Malawi are used in this chapter as cases to illustrate issues related to technology and technical education in developing countries. Over the past two decades, many countries have reformed their school curricula to establish technology as a key learning area for reasons that include the technological nature of society, national economic drivers, enhancing the opportunities of the disadvantaged, and possibilities for developing higher cognitive skills, including creative thinking and problem solving. Implementing significant school change is, however, complex and costly. While the rhetoric of *Technology Education for All* in the global north has been to distinguish it from vocational education, in Bangladesh, Malawi and other emergent economies, the relevance of education to everyday life is paramount. In these countries, a vocational emphasis might mean that a greater proportion of the population attend school if its usefulness and relevance is more obvious to students and their families.

## Introduction

This chapter considers the broad technology and technical education of two developing countries, Bangladesh and Malawi. These countries were chosen as they are both ex-British colonies, current members of the Commonwealth, share similar logistical and economic difficulties in relation to teachers and teaching, but are very different in their geography and language of instruction. Both have severe

F. Banks (✉)
The Open University, Milton Keynes, UK
e-mail: frank.banks@open.ac.uk

V.K.M. Chikasanda
University of Malawi, Zomba, Malawi
e-mail: vchikasanda@poly.ac.mw

 
PJ. Williams et al. (eds.), *The Future of Technology Education*, Contemporary Issues in Technology Education, DOI 10.1007/978-981-287-170-1_12

problems with school student completion rates. Both have aspirations to increase their economic base and look to technical subjects to supply the necessary expertise, but neither has the subject 'Technology' as part of general education within the curriculum. Frank Banks, with indigenous graduate students, has spent many years as an 'outsider' observing classrooms in Bangladesh. Vanwyk Chikasanda has conducted studies as an 'insider' familiar with the culture and pedagogy of Malawi. In looking to the future we speculate that both in its content and its pedagogy, technology as a school subject could be a cornerstone in providing a relevant curriculum for all, which could help address the issues of poor school attendance and the future needs of these countries.

Over the past two decades, many countries have reformed their school curricula to establish technology as a learning area (de Vries 2006). However, research (Dakers 2006; Dugger 2006) shows that most countries that have adopted technology education are from the western bloc of the world economy, with advanced technological developments. In the West, the rationale for technology education was driven by economic, social and educational assumptions, but technological advancement also influenced the nature of the technology education (Lewis 2000). Many other countries, particularly those categorised as the least developed, have maintained their imperial curriculum as they have been preoccupied with democratisation, poverty alleviation and other socio-economic problems. For example, Kerre (1994) attributed the poor state of technology education in least developed countries such as Bangladesh and Malawi to political instability, resource constraints and lack of educational leaders with a shared understanding of technology.

In 2006, when technology education was introduced into the curriculum of the Republic of South Africa, the reasons for doing so included *enhancing the opportunities of the disadvantaged*, the technological nature of society, *national economic problems*, possibilities for personal development in the higher cognitive skills, and *creative thinking and problem solving* (our emphases). The World Bank (2006a) and other agencies have identified similar needs in Bangladesh and Malawi, and in many parts of the world the rationale for technology education sits on a spectrum from 'development in higher cognitive skills, creativity and problem solving' to 'vocational preparation and work-based skills'. Among the sub-Saharan African countries, South Africa and Botswana have incorporated technology education and design and technology respectively as learning areas in their curricula (Stevens 2006; Weeks 2005). These two sub-Saharan nations have a much larger and better developed capacity for investment in education compared to other countries within the Southern African region, and their positions influence knowledge and policy developments in the region.

Any change in curriculum, whatever its rationale, can only be effective if students are present in school, and drop out and poor school attendance is a problem in both Bangladesh and Malawi. This was brought home starkly to Frank when he was working in Bangladesh:

> I was having a short break from a teacher education project that I was working on in Bangladesh and, with a colleague, took a small boat on the river delta near the capital Dhaka to look at the remarkable country outside the big city. The boat eventually put in to a

village and I fell into conversation with the head teacher of the school, who told me about the numbers of students on the roll, the numbers of classes at the school and the number of teaching shifts for each day. My mental calculation worked out that the class I was invited to see should have about 50 students—but there were just 5 attending on that day. I wondered why that was. Why were the students not coming to school?

Research suggests that there are broadly four reasons for poor school attendance in developing countries (Banks 2011). First, poverty is given as a principal reason, and is no doubt a key element—extra pairs of hands are needed at crucial times, such as harvest, and to look after younger siblings at home so that parents can work. But this is not the only, nor is it the major reason. Probably three more significant reasons are related to assessment, pedagogy and curriculum. The assessment regime is one that tests simple recall of facts rather than processes such as problem-solving that are useful in everyday life; the teaching is often uninspiring and, the students would say, too often teachers use corporal punishment; but probably most significant is that much of the school curriculum is considered irrelevant by both students and parents (Shohel and Banks 2010). The relevance of the curriculum—the ways in which what is studied in school has any relevance to the day to day life and needs of the students in their community—is of key importance to developing economies and we argue that the inclusion of technology as a school subject in countries like Bangladesh and Malawi could not only provide relevant content but could also help move the curriculum away from one based on memorisation of facts to one that provides useful process skills for life.

## Bangladesh and Malawi—The Contexts

Bangladesh is a semi-tropical country situated in the north-eastern region of South Asia, bordered by India and Myanmar (Burma), and is one of the largest deltas in the world. Its land is consequently very low-lying and crossed by three great rivers—the Ganges, Brahmaputra, and Meghna Rivers all flow south into the Bay of Bengal and their many tributaries make travel within Bangladesh difficult. Bangladesh is the eighth most populous country in the world and one of the most densely populated, with a population of 138.6 million crowded into an area of only 147,570 km$^2$ and a population density of 926 people per square kilometre. The Netherlands, in contrast, has similar topography but approximately 400 people per square kilometre. Over three quarters of Bangladeshi people live in the rural areas. Nearly half the population is under 19 years of age (MoPME 2008) and 76 % of the population live on less than $2 per day.

Primary education is provided to children from 6 to 10 years of age, in Classes 1 to 5, with an examination at the end of each academic year. In 2005, the Department of Primary Education in Bangladesh conducted a survey, which revealed that 47 % of students do not complete primary school, ranging in different areas form 30 to 72 % (MoPME 2008). Secondary education occurs through Classes 6 to 10, divided into two groups: Classes 6 to 8 form the Lower Secondary level, with a terminal examination; Upper Secondary is Classes 9 and 10, with the public Secondary School Certificate (SSC) examination conducted at the end of Class 10. Higher Secondary Education comprises Classes 11 and 12, with the public Higher School Certificate (HSC) examination taken at the end of Class 12. At either Class 8 or 10 students can choose to go into vocational streams (usually vocational education), or stay on and complete the general education stream. The choice of moving to the vocational education stream is voluntary. There are about 18 million students at the primary level and about eight million in secondary education. Surprisingly, although almost all primary schools are government controlled, of the 18,500 secondary level institutions in Bangladesh (excluding the Madrasas faith-based schools), less than 2 % (317) are government secondary schools. Nearly all secondary schools are private, although the government through the examination system specifies the syllabus and also pays teachers' stipends. Class sizes are large, with 60–90 students not uncommon in primary classrooms in urban areas.

Malawi is a landlocked country situated in Southern Africa with a population of 13.1 million, of which about 52 % is under the age of 18 (National Statistical Office of Malawi 2008). Like Bangladesh, Malawi therefore has a young population that needs sound education programmes if its society is to achieve technological literacy and leapfrog from the poverty cocoon. Malawi is the 17th poorest country in the world, with a Human Development Index (HDI) of 0.437 while Bangladesh is the 41st poorest with an HDI of 0.515 (United Nations Development Programme 2013). According to an *Integrated Household Survey 2004/05* and the *2006 Millennium Development Goals (MDG) Malawi* report, the current poverty estimates place 52.4 % of the Malawi's population below the poverty line, with 22 % in ultra-poverty (Ministry of Finance 2006). Malawi's poverty is attributed to limited access to land, low education, poor health status, limited off-farm employment, low technological developments and lack of access to credit, which has resulted from poor social, human capital and economic indicators (Ministry of Economic Planning and Development 2004; Ministry of Finance and Economic Planning 2002). In order to address the gaps, the *Malawi Growth and Development Strategy II* recommends a shift from an importing and consuming country to a producing and exporting country (Ministry of Economic Planning and Development 2012) where the role of science and technology is indispensable.

Malawi's economy is dependent on rain-fed agriculture and tobacco, with sugar and coffee accounting for 60 % of Malawi's export earnings. Over the last decade, the Malawi Government has emphasised diversification of the economic base through manufacturing and adding value to agriculture products (Ministry of Finance and Economic Planning 2002; National Economic Council 2003). As the current drive is for a science and technology-led economy, technology as a school subject is well placed to popularise societal technology and dispose populaces towards its acceptance and creation (National Research Council of Malawi 2002). The labour market in Malawi is very small due to the lack of a strong industrial and manufacturing base. Manufacturing accounts for 22 % of GDP while the agriculture sector accounts for 27 % of Malawi's GDP (Ministry of Finance 2006).

About 90 % of Malawi's population lives in rural areas and often rely on indigenous farming techniques (Ministry of Finance and Economic Planning 2002).

Technology education has potential to bring new perspectives towards rural development, which is an agenda of governments in Malawi, Bangladesh and most third world countries (World Bank 2006b) where the need for technology education is more compelling than it is in the more developed world. There are many situations that would change if there were 'technological minds' in the community. For instance, a primary school dropout from a rural village in Malawi's Dowa District developed a windmill to pump water and electrify his parents' home (see Kamkwamba and Mealer 2010). Another boy developed a radio station and was broadcasting to rural areas in the Mulanje District with messages that addressed rural needs. Yet another innovation arising from the need to address rural life problems, a villager developed a hydro power plant to drive a maize mill. While such innovations arise from some technologically minded youths, the lack of relevance of the curriculum to address the needs of living in a rural setting is significant.

The structure of education in Malawi is based on an 8 + 4 + 4 system: 8 years of primary school, 4 years secondary and 4 plus years of tertiary education. In 1994, Malawi introduced free primary education (FPE) in order to meet targets of Education for All and providing universal primary education, one of the targets of the Millennium Development Goals (Ministry of Education and Vocational Training 2000; World Bank 2004). With FPE, enrolment increased from 1.9 million in the 1993/1994 academic year to 3.2 million by 2000, creating pressure on school infrastructure, teachers, and teaching and learning materials. For example, the introduction of FPE increased the student/teacher and student/classroom ratios, estimated at 84:1 and 107:1 (Ministry of Education and Vocational Training 2006). FPE also increased competition for places in secondary schools and tertiary colleges, and the drop-out rate in primary schools increased significantly. According to the World Bank (2004), about 60 % of students drop out at the end of standard 8 (Year 8), with only 4 % proceeding to university after secondary school. It is indicated in the 2001 *Policy and Investment Framework* (PIF) that universal primary education gives the highest social returns on investment—a more economically active, informed, healthier and participatory population (Ministry of Education and Vocational Training 2000). Integrating programmes that promote technological literacy of citizens may help empower them with capabilities necessary for their meaningful contribution to national goals. FPE would also help Malawi increase the population literacy rate, which was estimated at 64 % (75 % male and 52 % female) during the 2004–2005 Integrated Household Survey.

Malawi's primary school curriculum has undergone various reforms in terms of subject content, pedagogy and assessment. Reforms in 2001, called the *Primary Curriculum and Assessment Reforms* (PCAR), were designed to address major shortfalls in educational attainment by primary school children while at the same time addressing national policy initiatives such as the *Malawi Poverty Reduction Strategy Paper* (MPRSP), *Vision 2020* and the 2001 *Policy and Investment Framework* (PIF). However, despite reviews to address emerging issues, the primary school curriculum has remained rather abstract, making it difficult for students to be able to transfer classroom activities to real life work. For example, most curricula promote academic reading, where transferability of knowledge to social and practical situations is unlikely at lower levels of schooling. With poverty, a high drop-out rate, high unemployment and limited opportunities for secondary education, the curriculum should empower the students for gainful activities beyond school. The review of the primary curriculum that took place in the 1990s added Creative Arts as a subject, but an appropriate teacher professional development programme was not put in place. Creative Arts included carpentry, building, pottery, tinsmith and welding as learning units, and teachers were required to have all these skills or utilise local artisans. The curriculum was very content heavy and required significant human and material resource development and investment. The challenges to implementation were compounded by the fact that the subject was not examinable and therefore teachers preferred teaching subjects in which their students would sit national examinations.

Both Bangladesh and Malawi, therefore, share similar concerns:

- School attendance is poor and drop out is high;
- The population is predominately young and potentially very economically productive;
- The curriculum and assessment systems are 'traditional', drawing largely on an imperialist past which has emphasised abstract knowledge, memorisation and recall;
- Education competes with other priorities for scarce resources;
- Educational reforms are decided centrally but implementation, especially in rural areas, is problematic due to poor infrastructure.
- Economic reforms emphasise a shift from a rural economy to one that focuses on manufacturing and science and technology-led development, making technology education relevant to both countries
- A curriculum that incorporates technology education may help foster both Bangladesh and Malawi's socio-economic development, as even a rudimentary understanding of technology enables one to evaluate, select and make more effective decisions regarding the use of technological products and services (Faure et al. 1972).

## Technical Education

### *Bangladesh*

In Bangladesh, technical streams (but not technology as a general subject) are available at both Secondary and Higher Secondary levels. The Bangladesh Technical Education Board (BTEB) specifies the curriculum, and technical streams are available from Classes 9 and 11. Although the National Curriculum in Bangladesh is specified through textbooks and examination syllabuses, a broad range of technical areas is available. For example, through the BTEB, students in Class 9 in a technical stream could be offered Automotive, Wood Working, Dress Making and Tailoring/ Garments Manufacturing, Fish Culture and Breeding, Fruit and Vegetable

Cultivation, Plumbing and Pipe Fitting, and Industrial Electronics (BTEB 2013). However, it is highly unlikely that a secondary school, particularly in rural areas, has the resources to offer these subjects in the necessary practical way. The Campaign for Popular Education, Bangladesh (CAMPE 2008) notes:

> A high degree of inequity exists in the secondary education sub-sector in Bangladesh. Inequity starts with unequal distribution of basic school facilities. All types of secondary educational institutions lack basic minimum requirements for quality education. [...] As learning performance in secondary education has direct implications for future life, the above inequities persist throughout the life of the secondary graduates, afflicting adversely their further education and employment opportunities. (p. xxxiii)

This lack of resources leads to an overly theoretical approach to technical education, which is endemic in Bangladesh. The country is also saddled with an examination system that rewards the memorisation of facts at the expense of opportunities for problem solving and critical enquiry.

In 2009, Frank initiated a large-scale study of general pedagogical practices in Bangladeshi classrooms (EIA 2009). A total of 252 classroom observations were undertaken in both primary and secondary schools. A 'time sampling' technique was used to record what type of activity (from a pre-determined list) the teacher and students were doing as the lesson progressed. The observers could also annotate the instrument with details that would complete the account of the lesson. The data provided an indication of the types of activity that happen in classes at the start, during and at the end of lessons. Across the lessons, teaching from the blackboard or front of the class was the predominant pedagogic approach. As the lesson progressed, teachers tended to read from the textbook, ask closed questions or move around the classroom monitoring and facilitating students' individual learning activities. The use of teaching aids (other than the textbook) was infrequently observed: between 2 and 6 % of classes at any of the times sampled. More frequently, teachers gave instructions for student activities (from 5 to 8 % at any of the times sampled) or listened to students as they read aloud from the textbook (from 2 to 8 % at any of the times sampled). At the end of a lesson teachers usually assigned homework (53 % of classes) and/or summarised what the lesson has just covered (49 % of classes). In many cases teachers provided feedback on the students' performance throughout the lesson (43 %) and assessed students' understanding by asking summary questions (34 %). In almost 10 % of the lessons observed, the teacher simply stopped teaching and left the room.

In the EIA (2009) study, the majority of teachers appeared to be fully or partially confident with the subject matter of the lesson. Teachers with a general training in education appeared to be more confident than others. However, there was little evidence of a lesson plan being used by teachers—only 14 % did so either 'regularly' or 'occasionally'. Most teachers interacted positively with their students and maintained good discipline when being observed. Few teachers focused their attention only on those students at the front of the classroom (8 %) while the majority focused on students throughout the class. However, most teachers did not adopt a stimulating and task-based approach to their lessons. Overall, 58 % did not ask any thoughtful questions to stimulate students' interest and 48 % did not set any challenging tasks for the students to make them think.

Follow-up studies adopting more ethnographic approaches were able to probe behind the direct observations of *ad hoc* visiting researchers. The following are translated quotations from interviews conducted with pupils.

> Almost every day teachers give us homework. Most of it memorising answers to the selected question. It's very hard for me to cope with the homework load. I'm afraid of being punished in the classroom. When I don't prepare my homework, I start to pray to God silently in the bad tempered teachers' classes. However, sometimes I was asked for homework and beaten. [Student Grade-VI]

> Though we're learning lots from secondary school, I'm not sure how much would be useful for our life, especially if we can't carry on after secondary school. I think it could be better if we learn something which will help us to earn some money and make our life a bit easier. I don't know what could be done for us. But we really need something which could make our lives comfortable and enjoyable. [Student, Grade-VIII] (Shohel 2010, p. 30)

Our observations of the classrooms showed that in most classes, students were not interactive at all; rather they were very passive learners. They only participated by answering the questions asked by the teacher. Generally the students were well behaved in class and in the majority of classes there were few students who had problems concentrating and/or displaying inappropriate behaviour. They were generally inactive and bored, and the curriculum favoured an abstract approach that prized memorisation over practical ability—something highlighted when Bangladesh vocational education was contrasted with Germany (Ahmed 2010). Moreover, the school environment was uninspiring. Although classrooms were generally clean and tidy, with good natural light and basic teaching equipment like a blackboard and chalk, and sufficient furniture for the students present in class, there was little evidence of students' work on display and different learning and teaching materials were not often used. The predominance of memorisation for a knowledge recall examination and extensive content coverage is not highly successful—pass rates of the Secondary School Certificate (SSC) are about 60 % (MoPME 2008).

Although it has an economic growth rate of over 5 %, Bangladesh will need to create at least 2.25 million jobs per year to accommodate a near doubling of the labour force from its present size of 55 million to 100 million in 2020. Technical and Vocational Education (TVE) is provided through government, non-government and the Bangladesh Technical Education board (BTEB) certified institutions around the country. As indicated above, students interested in pursuing TVE have the opportunity to enrol in government/non-government technical and vocational institutes after completing the junior-secondary level (Classes 6 to 8). The National Skill Level for Class 9 completion in TVET is 2. The school leaving qualification is Secondary School Certificate (SSC) Vocational (National Skill Level – 3) is equivalent to the general SSC. At the intermediate level there is the Higher Secondary Certificate (HSC) Vocational (National Skill Level – 4), which is also equivalent to the general HSC. At the post-secondary level, an individual can enrol at a tertiary education institution for an advanced degree, or a training institution for a diploma. Students graduating from both general SSC and SSC Vocational can enrol in government and non-government Polytechnic institutes for a Diploma in Engineering. These graduates also have the opportunity to study for a BSc in Engineering from engineering universities around the country, but the places are limited.

The BTEB-certified institutions provide training in different trades including, for example, ICT and medical technology. There is also the National Youth Development and Self-employment Academy (NYDASA), a government-certified, privately-run institution providing training in medical technology, ICT, entrepreneurship, communication and so forth. However, the emphasis on abstract knowledge and memorisation of information permeates even this high-level engineering curriculum. For example, in a comparison of electrical engineering course curricula and student practical competence between Bangladeshi and German students, Ahmed (2010) noted:

> The curriculum for the Diploma-in-Engineering (Electronics Technology) has been analysed. The outcome (students' competency level), particularly in the case of application oriented tasks, was measured through a competence test. In this competence test [Bangladeshi] polytechnic students came off badly and lag far behind the vocational school trainees in Germany. A comparison of the findings in Bangladesh and Germany, as presented in this research, reveals that the difference in the students' performances can be explained by the differences in the two countries' curricular areas of emphasis and different focuses in their respective curricula. The Diploma-in-Engineering curriculum in Bangladesh covers a broad spectrum of curriculum content and focuses mainly on theoretical matters. In Germany the curriculum is relatively specialised and it emphasises practical tasks. (p. 149)

The World Bank (2006a) has similarly noted that an emphasis on examinations has a detrimental effect on the quality of technical and vocation education and that:

> Institutions [in Bangladesh] suffer from under-utilization of resources, lack of equipment, unavailability of qualified instructors, low levels of enrolment, high drop-out rates, shortages of teachers' training facilities, and a high degree of centralization. The lack of resources and under-utilization of those that are available is indicative of poor distribution and management of resources. Students often cannot participate in practicums, for example, and are forced to observe due to a lack of sufficient equipment. (p. 35)

## *Malawi*

In Malawi, as in Bangladesh and most other British colonies, the integration of vocational and liberal education has largely been based on the system prevalent in the UK at the time, and also heavily influenced by donor agencies. Currently, Malawi has over 700 secondary schools. However, through funding from the International Development Association (IDA), introduced technical subjects in 13 pilot secondary schools soon after gaining political independence from Britain in 1964. The aim was to attract able students into engineering-related study, as well as to equip students with skills for jobs and self-employment. The pilot was largely based on the premise that a comprehensive curriculum would help in adaptation of capabilities and increase social and occupational mobility (Urevbu 1988). Despite professional development initiatives undertaken in the early 1970s, the curriculum did not reflect the vocational needs and cultural contexts of rural communities and implementation remained only in the 13 pilot schools, which were given heavy and sophisticated machinery.

Malawi's indigenous technologies, culture, values and beliefs are not reflected in the curriculum. Further, real-life examples and contexts must be relevant or students will not easily transfer skills learnt in an industrial-based classroom to their local context since there are no industrial-related activities in rural villages, where most students are from. Although opponents of a vocational curriculum argued that the curriculum inhibits further study and thus reduces future socioeconomic attainment, proponents have argued that it helps students avoid unemployment and increases their chances of becoming skilled workers (Arum and Shavit 1995).

Many countries in Africa hoped that diversification of the curriculum would ease unemployment problems and significantly promote growth and productivity. However, poverty and youth unemployment have continued to escalate despite the promises of such a curriculum. Student enrolment numbers for technical subjects have continued to dwindle and some schools have completely closed down their technical wings. Although policy and political will may overturn events, the curriculum still needs to be reviewed. Malawi needs a curriculum that promotes social, economic and environmental awareness and development, and at the same time enhances the beliefs and values of the society.

Malawi, like many African countries, currently offers the traditional technical subjects of metalwork, woodwork and technical drawing but teaching emphasises a narrow craft skills approach. Despite overarching education reforms in England (Banks and McCormick 2006), Malawi has maintained the traditional system with few subject changes. In particular, the craft and skills-based technical subjects in the general education curriculum have largely remained the same. Although Malawi is still a third world country, with no technological commonplaces (Lewis 2000), there is an awareness of the global trend to re-shape technical subjects towards technology. The *Malawi Poverty Reduction Strategy Paper* (MPRSP) (Ministry of Finance and Economic Planning 2002) recognised that: "The low content of science and technology in national economic development programmes is a barrier to economic growth leading to high levels of poverty among Malawians" (p. 92).

As the economy is agri-based, the *Malawi Growth and Development Strategy II* (Ministry of Economic Planning and Development 2012) focuses on reducing poverty through increased access to basic social services, accelerating growth and improving productivity in agriculture and the manufacturing sectors. Technology education, with appropriate pedagogy, could be critical in enhancing the technological capabilities of citizens for their participation in the country's economic activities. However, despite undertaking a number of national curriculum reforms (Nyirenda 2005), technical subjects in Malawi have not been reviewed to articulate emerging issues, government policies and development agendas. Malawi's *Vision 2020* statement stipulates that the nation aims for a technologically driven economy (National Economic Council 2003). To achieve this, *Vision 2020* recommended a review of the school curriculum, the promotion of skills training and the development and introduction of a culture of science and technology. The suggested curriculum reforms focus on national educational programmes that are more reflective of changing socio-economic and political realities. The strategic goals for attaining the Vision included strengthening science and technology education through the teaching of science in primary and secondary schools, as well as strengthening the teaching of computer studies and technical subjects. Earlier, the *Malawi Poverty Reduction Strategy*, a medium-term action plan, followed by the *Malawi Growth and Development Strategy* (Ministry of Finance and Economic Planning 2002), stipulated an intensified application of science and technology to be facilitated by the creation of a science and technology culture to encourage appreciation by the Malawian society for science and technology-led development. The *Science and Technology Policy for Malawi* also included a strategy to enhance technological literacy through curriculum changes to ensure an effective science and technology education and culture at all levels of the education system (National Research Council of Malawi 2002).

All major policy guidelines for Malawi (e.g. Vision 2020, PIF, the MPRSP, Millennium Development Goals and the National Science and Technology Policy) therefore place education at the fore of developing science and technology and eradicating poverty. However, as in Bangladesh, the current curriculum does not provide students with skills to become economically active. Therefore, those who drop out have difficulty finding gainful employment or self-employment, let alone understanding the technological developments taking place, or that need to be undertaken, at the personal, community or national level. Five decades after independence, the curriculum remains much the same. In its current form, it provides little scope for developing student capabilities so they can understand, create, control and manipulate technology. In order to address the policy strategies for technological literacy, computer studies and science and technology were introduced as learning areas in 2001 (Ministry of Education and Vocational Training 2001a, b), but curriculum goals and objectives are similar to the assumptions guiding curriculum vocationalisation, which focussed on craft and artisan skills development for employment. Science and technology was also established as a core learning area to provide learners with an understanding of the close relationship between scientific knowledge and technological applications.

As a means of attaining relevance in the curriculum, the *Policy and Investment Framework* (PIF) (Ministry of Education and Vocational Training 2000) stipulated that:

> …the primary and secondary school curriculum of the future should strive to impart essential skills and knowledge on a broad range of issues including new basic skills: critical thinking and analytical skills, civic and democratic values, computer skills, entrepreneurial skills, life skills and environmental education. (p. 12)

As a consequence, the *Malawi Poverty Reduction Strategy Paper* (MPRSP) stipulated an intensified application of science and technology to be facilitated by the creation of an S&T culture in order to foster its appreciation in society. The S&T policy for Malawi also included, as a strategy, the upgrading of the S&T curriculum to enhance technological literacy (National Research Council of Malawi 2002). In response to these policy guidelines, new syllabi were developed in 2000 (Ministry of Education and Vocational Training 2001b). These incorporated Science and Technology in the school curriculum as an integrated core learning area, replacing Physics and General Science. The Science and Technology syllabus included such broad topics as planning and performing scientific investigations, properties and uses of matter, energy forms and conversions, environment, population, introduction to technology, biotechnology and indigenous and industrial technologies. After a protracted debate with the University of Malawi over the content and relevance of the new subject, Physics and General Science subjects were re-introduced and became core subjects again as they are prerequisites for entry to many science-based university programmes in Malawi. During the development of the new syllabi, technical subjects were not reviewed and have continued to be offered independent from, and alongside, Science and Technology. It was therefore evident that the technical education curriculum was side-lined in the implementation of policy strategies. This was despite its perceived prominence in, and centrality for, harnessing Malawi's technological development potential and poverty alleviation initiatives. For instance, among several strategies for improving prospects for economic growth, the introduction of vocational, technical and business management courses at primary and secondary schools were recommended as strategies for addressing gaps in the human resource base (Ministry of Economic Planning and Development 2004).

Malawi therefore recognises the importance of technical, entrepreneurial and vocational education and training (TEVET) for sustainable economic growth and development. However, the technical education curriculum needs to be reviewed so as to offer students broad-based technology education. Students' experiences in technology education help enhance their capabilities to be able to fully participate in society initiatives for a technologically driven economy (Jones 2003). The *Malawi Growth and Development Strategy* (MGDS) included reviewing and reforming school curricula to address national needs as a key strategy for enhancing the quality and relevance of education. An alternative teaching and learning approach emphasising process skills and problem solving may enhance students' technological literacy in line with the government's development agenda as stipulated in its major policies. Government recognises manufacturing as key to economic growth, however, and aims to improve the quality of products and productivity of both labour and capital and enhance human capital through better integration of science and technology into vocational training and improving standard certification capacity.

In practice, the strategies for addressing a technology culture as outlined in *Vision 2020* appear to have been interpreted in the same manner as the assumptions of a vocationalised curriculum. For instance, the goal for computer studies appeared to focus on realigning learners with computer-related jobs. The introduction of the new subjects therefore appeared to be an attempt to de-establish technical subjects when policy guidelines still demanded an enhancement of technological literacy. Hence, there is a need to review the technical curriculum so that there are opportunities to provide an alternative education that can impart knowledge, capabilities and skills responsive to the social, economic, and environmental climate of Malawi.

## Implementing a Relevant Curriculum—Some Examples

### *Bangladesh*

As in Malawi, assumptions for technology education in Bangladesh are rooted in assumptions guiding vocationalisation. However, it is important—as illustrated above—to consider the perceived needs of both students and their parents and what they consider relevant to be learned and what will encourage them to appreciate the benefits of schooling. This is particularly so for students from poor families. The Underprivileged Children Education Programme (UCEP 2013) in Bangladesh provides general education and vocational training for over 30,000 poor working children who have generally missed out on their primary education. The children continue to work and earn while they attend school. To enable this, UCEP schools operate three shifts per day, each of 3 h duration. A child chooses a shift of his/her convenience, in consultation with parents (guardians), to minimise the economic loss to the family when the child attends school. The schools offer the standard national curriculum but taught over a shorter period; each year's syllabus is completed over a 6-month period using the curriculum and textbooks prescribed by the National Curriculum and Textbook Board (NCTB). Basic elements of technical education are also included. At UCEP schools attendance is around 94 %, in marked contrast to government schools.

Similarly, schools run by Bangladesh Rural Advancement Committee (BRAC) for the rural poor offer informal education to many children:

> Most BRAC nonformal schools are one-room schools with limited floor space. The classroom is very neat and clean and students sit on mats on the floor. There are commonly about 30 students, two thirds of whom are usually girls. The teacher, generally a female with at least 10 years of schooling, is chosen from the community where the school is situated. [...] The informality of the nonformal school environment has flexibility for teaching and learning in a community context where interactions between school and community are very influential and fruitful for students' development. (Shohel and Howes 2008, pp. 293–294)

The difference in attendance and the perceived curriculum relevance of these non-government schools is stark. BRAC schools link the curriculum to basic hygiene and health education, whereas little is done in government schools to consider the usefulness of science or technology to agriculture or the need for clean food and water. For example, Shohel and Howes (2008) tell of a school home economics lesson on the need for cleanliness in the home to prevent disease, taking place in a very dirty classroom.

UCEP schools offer an integrated general and vocational curriculum. Students generally follow the government-specified curriculum both at primary and lower secondary level (Classes 1 to 8). The curriculum consists of Bangla (the mother tongue), English, mathematics, vocational, social, environment and hygiene education. As the focus is to educate poor working children in urban environments who have never been enrolled at a government school, they are accepted into the programme no younger than age 10 for girls and 11 for boys. Each 3 h shift is focused on general education, but where possible examples are drawn from a technical or technological context. For example, the English alphabet is taught and illustrated through the naming of craft tools—D for dividers, H for hammer, and so on. Stories in Bangla are linked to the discovery of inventions and the use of agricultural and other devices. After Class 8, UCEP continues Technical Education training in 16 trades, including automechanics, electronic technology, industrial electrical and electronic control, offset printing technology, industrial woodworking, and tailoring and industrial sewing operation. Students learn in a highly vocational and practical way, using English where necessary for technical vocabulary (EIA 2009). At the end of their training they are guaranteed a job. In contrast to attendance in the formal government system, these poor working children attend school regularly and complete their education. Both students and parents see the curriculum as relevant and worthwhile, not only in terms of content but for imparting appropriate life and employment skills through an active learning pedagogy.

## *Malawi*

Technological literacy in Malawi is still a central part of the government's envisioned plans for technology-led development, suggesting a shift to a new curriculum model for student learning. Such a curriculum may need to empower students to think and reason through problem solving, leading to students acquiring capabilities necessary for self-reliance and confidence for effective participation in social and economic development. While the current context of high poverty, a small industrial base, and a predominantly rural economy may appear to hinder effective learning of technology, the same conditions may also be viewed as providing rich opportunities and contexts for students' meaningful learning, as the students can be challenged to address authentic issues affecting the communities in which they live.

In the primary school sector, 24 model schools were built in the late 1970s and supplied with equipment and primary school teachers for vocational training. The model schools offered vocational skills through a Craft and Technology curriculum that included aspects of content areas such as Tinsmith, Woodwork, Metalwork and Technical Drawing. Some of the teachers were trained through the Malawi Young Pioneers (MYP), a paramilitary wing of the then ruling Malawi Congress Party. The MYP was a youth scheme established to train boys and girls in various skills/ vocations like agriculture, building and construction, carpentry and joinery, and many more. The training was undertaken at various training bases, like Nasawa in Zomba, Kamwanjiwa in Mzimba, and many others. After training, the graduating youths were placed in various institutions but those that majored in agriculture were given land in settlement schemes that were established in various districts like the Chinguluwe settlement scheme in Salima district.

The paramilitary capacity of the MYP did not augur well with the constitutional obligations of the Malawi Armed Forces and towards the first multiparty democratic presidential and parliamentary elections in 1994 the group was disbanded, leading to closure of all MYP training institutions, which impacted on teaching and learning in the model primary schools. The Ministry of Education, Science and Technology has re-introduced technology studies and a pilot project is being implemented at Senga Model Primary School. An evaluation of the pilot programme was planned in order to understand the strengths and challenges of the programme and the ability of the programme to meet its goals and to inform the "rolling out" of the programme to other model schools in the future.

Technical education at tertiary level in Malawi was established in 1965 after the opening of the Polytechnic as a constituent college of the University of Malawi. The Polytechnic, with a mandate to offer both technician-level and tertiary-level courses, was able to offer tertiary technical courses at diploma and degree levels in engineering, construction, business studies, commerce, laboratory techniques, public health inspection and secondary school technical teaching. Technician apprenticeship courses were offered in motor vehicle mechanics, general fitting and electricity, while craft level courses were offered at the technical schools. Technical schools have continued to offer artisan training but technician courses at the Polytechnic were discontinued in 2002 with an aim of upgrading technical schools to begin offering technician-level programmes. However, the colleges have not been upgraded due to capacity deficiencies—leading to large gaps in the technician to engineer ratio in industry, impacting on manufacturing.

Following *Vision 2020* and a 1996 Gesellschaft für Technische Zusammenarbeit (GTZ) sector study, a TEVET policy was developed. The policy focused on strategies outlined in *Vision 2020,* which centred on promoting sustainable development and poverty reduction, including diversifying the economy through industrialisation (National Economic Council 2003). Through reforms that established Competency Based Education (CBE) programmes and the TEVET Qualifications Framework (TQF), greater emphasis was placed on appropriate and demand-driven technical, vocational and entrepreneurship education and training. The system has attempted to meet the expectations of industries but the extent of the delivery of the programmes is impacted by limitations of classroom space, outdated curriculum and limited offerings, staff capacity gaps in colleges and industry, and technology development challenges. Although the TQF provides national standards for learners, trainers and employers in the TEVET system, its roll-out has faced more challenges than anticipated. Regular reviews are necessary to inform best practice in the system.

Despite such progress in TEVET activities, there is no comparable change in economic growth that can be attributed to TEVET's role in the sector. The TEVET system therefore needs further research and reforms for it to contribute meaningfully to rapid and sustainable poverty alleviation of Malawians living below the poverty threshold. While the formal training programmes are worthwhile for the industrial and manufacturing sectors of the economy, the informal sector needs support as this is where a large number of unemployed youths would get jobs and services.

The informal sector involves small-scale business ventures that employ a large section of the untrained workforce. Activities are wide-ranging: merchandise, such as buying and selling of commodities to get a profit; production of goods such as furniture, mats, clothes; growing of crops, raising of animals; and services such as maintenance of domestic equipment, cars or bicycles. These are common in slum areas such as Kawale in Lilongwe and Ndirande in Blantyre. Mostly such businesses are difficult to track by regulatory and tax collecting bodies. However, such businesses need support in terms of capital equipment, business skills and technology improvement—which could be provided within a targeted and coordinated TEVET policy framework. Further, TEVET reforms should be enacted in a consolidated manner in order to develop a coherent and unified framework that addresses the needs of all social groups and provides flexible, sustainable, gender-neutral pathways. Reforms should also be informed by locally derived research-based evidence rather than the wholesale adoption of philosophies and theories from countries with very different educational contexts.

In both Malawi and Bangladesh, therefore, the current school curriculum does not provide students with skills to become economically active. Despite the political rhetoric, the curriculum is abstract and semi-detached from the daily realities of students. Those who drop out have difficulty getting jobs or becoming self-employed, let alone understanding the technological developments taking place, or that need to be undertaken, at the personal, community or national level. In both countries there is a need to review the curriculum to include technology education so that there are opportunities to provide an alternative education—one that is not based on memory and the recall of abstract facts, but that that can impart knowledge, capabilities and skills responsive to the social, economic, and environmental climate of both Bangladesh and Malawi.

## Looking to the Future

Having set out the current situations in Malawi and Bangladesh, what could a general technology education offer students in these countries and what lessons have been learned about its possible implementation?

The evidence from the UCEP model illustrates the over simplification of the view that school drop-out in countries such as Bangladesh or Malawi is solely due to poverty. It is certainly a factor, but research suggests that more significant is the need for a relevant curriculum that provides a meaningful purpose for students to attend school (Shohel 2010). In the case of UCEP, this purpose is provided through a technical curriculum. The very high attendance and low drop-out rate for very poor students at these schools support this view. Technological literacy at the primary level, particularly prominent in UCEP schools, and to some extent in BRAC schools, shows the need for stronger links between education and real life. Similarly, the experience of the MYP experiment of the early 1990s in Malawi and the aims of *Vision 2020* show

Table 12.1 SEL training methods (Quoted in Banks 1994, p. 206)

| Old | New |
| --- | --- |
| 'Show and Copy' | Projects |
| Fabrication of pre-prepared items | Work out what is needed to construct items |
| Superficial discussion | Simulation of construction, including estimates and other costs |
| *Teaching styles* | |
| Didactic | Learner centred |
| Respond to a brief | Respond to an identified need |
| Solitary learning | Cooperative learning |
| Copy isolated tasks | Total process |
| Facts based | Processes important |
| *Learning outcomes* | |
| Facts, specialised knowledge | Facts, methods, social skills, evaluative skills and an ability to cooperate |

how links between technology education, and technical and vocational education, can lead to education being seen as more than memorisation for an examination—by students, teachers *and* parents. There is also the need for a stimulating learning environment, with an increase in the use of student-centred learning techniques.

This chapter has taken the view that technology education can provide a relevant curriculum for all students that will help reduce poor attendance and high drop-out rates. The rhetoric of Technology Education for All in the global north has been to distinguish it from vocational education. In Bangladesh, Malawi and other emergent economies, however, the relevance of education to everyday life is paramount and a vocational emphasis might mean that a greater proportion of the population attend school as the usefulness would be more obvious to students and their families.

Friedrich Ebert, founder of the German Social Democratic Party, once said: "General education is the vocational education of the upper classes; vocational education is the general education of the working class" (in Finegold et al. 1990, p. 3). When primary education is the only education of the majority of students, an emphasis on examples drawn from a vocational context and a pedagogy that encourages active learning and process skills has many wider benefits. Indeed, the German communications firm SEL has moved to a vocational training course (see Table 12.1) that is very similar to the approach we are advocating.

Hassan (2013), working in collaboration with Frank, conducted a comprehensive 5-year study into difficulties implementing pedagogical change in rural Bangladesh. He set out the implications when enacting change for teachers, teacher educators, materials developers, and policy makers. His conclusions are adapted and abridged here:

## *Implications for Teachers*

If teachers are to follow a student-centred approach, they need to change their practice in the following key areas:

Pedagogy

- It is essential for teachers to do more than simply address the class timetable, lesson plans and preparation, and classroom management to ensure quality teaching;
- Teachers need to be familiar with and practise general teaching techniques like setting out beginning, middle and ending activities in lessons;
- Teachers should use teaching aids when necessary in the lesson to make learning interesting and participatory;
- Teachers should create opportunities for students to talk more so that they practise and improve communication skills. The teacher should talk less so that he/she organizes and facilitates students' activities in the lessons;
- Teachers should create a friendly and participatory environment to ensure everyone's participation in the class activities;
- Teachers should avoid offering private tuition so that they can give full attention to all students in their classes;
- Finally, teachers should focus on students' needs and their interests in the class.

## *Implications for Teacher Educators*

- Introduce appropriate courses for both pre-service and in-service education;
- Provide opportunities for workshops or seminars and involve teachers in policy making (see also Chap. 13, this volume, by Kendall Starkweather);
- Improve teachers' own subject proficiency.

## *Implications for Materials Developers*

- As teachers are generally lacking in opportunities to attend teacher training programmes, workshops or seminars, textbooks and the Teachers' Guide should help fill the gap and be widely disseminated;
- Materials developers should take teachers' subject proficiency into account when they compile teaching materials, thus enabling teachers to be better able to implement the methodologies. They should also guide teachers towards other relevant sources to widen the range of teaching materials.

## *Implications for Policy Makers*

The implementation of a new teaching approach is intended to meet a social need, and policy makers should be sensitive to what is happening in the global domain. However, this does not mean that they should blindly copy international trends. Any policy made should be practical and applicable locally, taking the needs of students and teachers into account.

With respect to students, policy makers should:

- strive to ensure a conducive and friendly learning environment in the schools;
- provide equal opportunities with respect to such aspects as teachers, teaching materials, and the teaching and learning environment for *all* students regardless of whether they live in urban or rural contexts; and
- consider the background of students and take necessary measures to ensure their general participation and learning in the schools.

With respect to teachers, policy makers should:

- take appropriate measures so that teachers have sufficient opportunities to receive adequate training;
- take into account teachers' professional conditions, such as recruitment and promotion processes;
- ensure training facilities for in-service to help ensure that innovations are sustainable by teachers;
- involve teachers in discussions when a new teaching methodology is implemented;
- ensure a standard teacher-student ratio so that teaching and learning can take place in a productive class-size environment;
- improve monitoring systems to ensure classes are adequately and appropriately conducted; and
- have an appropriate examination system with pedagogies relevant to its successful implementation. Policy makers should therefore consider the examination process when implementing a new teaching approach (see Chap. 7, this volume, by Kay Stables).

## Concluding Thoughts

It is evident from the examples above that the teacher education systems in Bangladesh and Malawi need to be revamped if they are to provide opportunities for further professional growth of teachers, and the development for improved teaching practices, ultimately leading to the development of professionals who can construct their own classroom research to test new innovations, theories and pedagogy. More research is also required to help develop the capacity for reconstructing technology education in Malawi and Bangladesh in ways that reflect a shift from an industrial arts-based curriculum to a broad-based technology education that is relevant to students' current and future lives.

Technology education in developing economies needs to be introduced as subject with the potential to facilitate major changes in general and vocational education, as it should be in western economic contexts. In the words of James Callaghan, a former British Prime Minister who witnessed the independence of both Bangladesh and Malawi:

> The goals of our education, from nursery school through to adult education are clear enough. They are to equip children to the best of their ability for a lively, constructive place in society and also to fit them to do a job of work. Not one or the other, but both. (Callaghan 1976, p. 332)

## References


Ahmed, Md. F. (2010). *Technical and Vocational Education and Training–curricula reform demand in Bangladesh. Qualification requirements, qualification deficits and reform perspectives*. Unpublished PhD dissertation, University of Stuttgart.

Arum, R., & Shavit, Y. (1995). Secondary vocational education and the transition from school to work. *Sociology of Education, 68*(3), 187–204.

Bangladesh Technical Education Board (BTEB). (2013). *BTEB Trade/Technology Info*. Accessed 20 Nov 2013 from: http://www.bteb.gov.bd/page.php?action=trade_technology

Banks, F. (1994). *Teaching technology*. London: Routledge.

Banks, F. (2011). Technological literacy in a developing world context: The case of Bangladesh. In M. de Vries (Ed.), *Positioning technology education in the curriculum* (pp. 219–225). Rotterdam: Sense.

Banks, F., & McCormick, R. (2006). A case study of the inter-relationship between science and technology: England 1984–2004. In M. J. de Vries & I. Mottier (Eds.), *International handbook of technology education: Reviewing the past twenty years* (pp. 285–311). Rotterdam: Sense.

Callaghan, J. (1976). Towards a national debate. Reprinted in *Education*, 22 October, 332–333.

Campaign for Popular Education (CAMPE). (2008). *The state of secondary education: Quality and equity challenges*. Dhaka: Author.

Dakers, J. R. (2006). Technology education in Scotland: An investigation of the past twenty years. In M. J. de Vries & I. Mottier (Eds.), *International handbook of technology education: Reviewing the past twenty years* (pp. 331–346). Rotterdam: Sense.

de Vries, M. J. (2006). Two decades of technology education in retrospect. In M. J. de Vries & I. Mottier (Eds.), *International handbook of technology education: Reviewing the past twenty years* (pp. 3–11). Rotterdam: Sense.

Dugger, W. E., Jr. (2006). Twenty years of educational standards for technology education in the United States. In M. J. de Vries & I. Mottier (Eds.), *International handbook of technology education: Reviewing the past twenty years* (pp. 65–81). Rotterdam: Sense.

English in Action. (EIA). (2009). *Baseline Study 3. An observation study of English lessons in primary and secondary schools in Bangladesh*. Milton Keynes: English in Action Monitoring and Evaluation Team.

Faure, E., Herrera, F., Kaddoura, A.-R., Lopes, H., Petrovcky, A. V., Rahnema, M., & Ward, F. C. (1972). *Learning to be: The world of education today and tomorrow*. Paris: UNESCO.

Finegold, D., Keep, E., Milibrand, D., Raff, D., Spours, K., & Young, M. (1990). *A British 'Baccalaureat'*. London: Institute for Public Policy Research.

Hassan, Md. K. (2013) *Teachers' and students' perceived difficulties in implementing communicative language teaching in Bangladesh: A critical study*. Unpublished PhD dissertation. The Open University, Milton Keynes, UK.

Jones, A. (2003). The development of a national curriculum in technology for New Zealand. *International Journal of Technology and Design Education, 13*(1), 83–99.

Kamkwamba, W., & Mealer, B. (2010). *The boy who harnessed the wind: Creating currents of electricity and hope (P.S.)*. Accessed 5 Nov 2013 from: http://www.williamkamkwamba.typepad.com/

Kerre, B. W. (1994). Technology education in Africa. In D. Layton (Ed.), *Innovations in science and technology education* (pp. 103–117). Paris: UNESCO.

Lewis, T. (2000). Technology education and developing countries. *International Journal of Technology and Design Education, 10*(3), 163–179.

Ministry of Economic Planning and Development. (2004). *Malawi economic growth strategy* (Vol. 2007). Lilongwe: Ministry of Economic Planning and Development.

Ministry of Economic Planning and Development. (2012). *Malawi Growth and Development Strategy II 2012–2016*. Lilongwe: Ministry of Economic Planning and Development.

Ministry of Education and Vocational Training. (2000). *Policy and Investment Framework*. Lilongwe: Ministry of Education.

Ministry of Education and Vocational Training. (2001a). *Senior secondary teaching syllabus: Computer studies, Forms 3–4*. Zomba: Malawi Institute of Education.

Ministry of Education and Vocational Training. (2001b). *Senior secondary teaching syllabus: Science and technology, Forms 3–4*. Domasi: Malawi Institute of Education.

Ministry of Education and Vocational Training. (2006). *Education statistics 2006*. Lilongwe: Ministry of Education and Vocational Training.

Ministry of Finance. (2006). *Malawi Growth and Development Strategy 2006–2011*. Lilongwe, Malawi: Ministry of Finance.

Ministry of Finance and Economic Planning. (2002). *Malawi poverty reduction strategy paper*. Lilongwe: Ministry of Finance and Economic Planning.

Ministry of Primary and Mass Education Government of Bangladesh (MoPME). (2008). *Supply, retention, preparation and career-long development of teachers. Bangladesh country paper*. Seventh E-9 Ministerial Review Meeting, Bali, 10–12 March.

National Economic Council. (2003). *Vision 2020: The national long term development perspective for Malawi*. Accessed 1 Mar 2014 from http://www.sdnp.org.mw/malawi/vision-2020/chapter-8.htm

National Research Council of Malawi. (2002). *Science and technology policy for Malawi*. Lilongwe: Office of the President and Cabinet.

National Statistical Office of Malawi. (2008). *Population and housing census preliminary results*. Accessed 1 Mar 2014 from: http://www.nsomalawi.mw/images/stories/data_on_line/demography/census_2008/Main%20Report/Census%20Main%20Report.pdf

Nyirenda, D. M. C. (2005, March-April). *Malawi secondary school curriculum reform: Issues and challenges*. Paper presented at the Malawi National Education Conference, Lilongwe, Malawi.

Shohel, M. M. C. (2010). *Transition from non-formal to formal education in Bangladesh: An exploration of the challenges students face*. Saarbrucken: LAMBERT Academic Publishing (LAP).

Shohel, M. M. C., & Banks, F. (2010). Teachers' professional development through EIA secondary teaching and learning programme in Bangladesh: Experience from the UCEP schools. *Procedia—Social and Behavioral Sciences Journal, 2*(2), 225–258.

Shohel, M. M. C., & Howes, A. J. (2008). Informality of teaching and learning in non-formal schools: Socio-cultural processes as mesosystems of student development. *Education from Age 3–13, 36*(3), 293–309.

Stevens, A. (2006). Technology teacher education in South Africa. In M. J. de Vries & I. Mottier (Eds.), *International handbook of technology education: Reviewing the past twenty years* (pp. 515–532). Rotterdam: Sense.

Underprivileged Children's Education Project (UCEP). (2013) *Technical Schools.* Accessed 20 Nov 2013 from: http://www.ucepbd.org/program/tech.htm

United Nations Development Programme (UNDP). (2013). *Human development report 2013.* New York: UNDP. Accessed 2 Nov 2013 from: http://hdr.undp.org/en/statistics/hdi/

Urevbu, A. O. (1988). Vocationalizing the secondary school curriculum: The African experience. *International Review of Education, 34*(2), 258–270.

Weeks, S. G. (2005). Prevocational secondary education in Botswana. In J. Lauglo & R. Maclean (Eds.), *Vocationalisation of secondary education revisited* (pp. 93–147). New York: Springer.

World Bank. (2004). *Cost, financing and school effectiveness of education in Malawi: A future of limited choices and endless opportunities.* Washington, DC: Development Research Group, The World Bank.

World Bank. (2006a). *The Bangladesh Vocational Education and Training system: An assessment.* Washington, DC: The World Bank.

World Bank. (2006b). *World development report 2007: Development and the next generation.* Washington, DC: The World Bank.

# Chapter 13
# Politics and Policy

**Kendall N. Starkweather**

*"Strive not to be a success, but rather to be of value."*
*Albert Einstein*

This chapter addresses the socio-political context of technology education from my background as a technology education professional followed by over three decades of experience as an executive with an international technology education association, the ITEEA. In order to be valued as a key area of learning, technology needs to be distinctive in the school curriculum and create a positive perception in the minds of parents and decision makers. This is an issue of branding. Technology education associations have a key role to play in positioning the subject, informing the politics and policy advancing the subject. Teachers who become active benefit from being involved in strategy discussions and resource development, networking advantages, and political support as a result of being known and interacting with others in the technology education and wider education communities. While much has been achieved by technology education associations, they will need to continue to evolve if they are to reflect contemporary values, beliefs and assumptions of the profession, and have robust mechanisms for supporting members to work together in a digital and globalised world.

## Introduction

Politics and policy will play an important role in the future of technology education, just as they have been a factor in many of the accomplishments that have been realised to date. Yet few in the profession fully recognise that in their efforts to have successful programmes, every attempt to should be made to show the value of a technology education. Technology educators often miss golden opportunities to promote their programmes and the profession through the accomplishments of their students as a result of high quality teaching and learning. This is important since it

K.N. Starkweather (✉)
International Technology and Engineering Educators Association (ITEEA), Reston, VA, USA
e-mail: knscls@icloud.com



PJ. Williams et al. (eds.), *The Future of Technology Education*, Contemporary Issues in Technology Education, DOI 10.1007/978-981-287-170-1_13

is when parents, school administrators, corporate decision makers, and elected policy makers raise their level of support for technology education that the field will have even greater opportunities to develop the next generation of technologists, innovators, designers, and engineers. These technological thinkers will be the valued leaders of tomorrow.

This chapter will address politics and policy from my background as a technology education professional followed by over three decades of experience as an executive with an international technology education association, the ITEEA (International Technology and Engineering Educators Association). Therefore, the content is generated from my experiences as well as more general association research. The intent of this chapter is to give the reader a perspective of the role of technology education associations in guiding the work of professionals, and how technology educators can use associations to advance their field through politics and policy. It is imperative that teachers be able to maximise their potential and know the "how and why" of being politically astute.

Advocacy by interest groups have resulted in the additions and changes that have taken place to curriculum over time. The advocacy of technology teaching continues as teachers communicate and share ideas about teaching and learning strategies and practices. Such advocacy groups often become associations that exist to help members share directions, set standards, and advance policy. However, the importance of politics and policy are seldom fully internalised by teachers, administrators and teacher educators, who each have a different primary concern: concentrating on teaching and learning.

This chapter will consider areas of politics and policy that have and will continue to affect the prominence of technology education. These include: positioning technology education, the power of a name, school politics, worldwide networking, association culture, core capabilities, forward thinking, raising the value of technology education, and showcasing success. My intention is that this chapter will help technology educators and members of technology education associations understand the important role of politics, be better advocates, and contribute to stronger policies for future technology education programmes.

## Positioning Technology Education

Technology is everywhere and can be linked to everything in our lives in terms of influencing our daily activities, the nature of our work patterns, the security of our information networks, and the safety of our countries. One would think that technology would be one of the most important subjects taught in our schools. However, it does not hold a key position in our schools today because of traditions that emphasise other subjects, understandings about the definition and philosophies for teaching about technology, and the ways that technology is learned in our daily lives or in schools. At a time when technology should be the most important subject in education, it is often considered an elective or 'add on' subject in a student's course of work.

The effort to keep technology education strongly positioned in the elementary and secondary school curriculum is ongoing. Technology education is not often considered as one of the primary core subjects with the status that is given to mathematics, reading, science, or social studies. In the United States, for example, technology education is not measured with the same consistency as the core subjects, if at all, signaling that it is not valued as highly. Rather, technology education is often thought of as a skills course having to do with how much one knows about and is able to use computers for learning, accessing information, or for pleasure—teaching technology is most often thought of in the form of teaching "with" technology rather than teaching "about" technology. For many in the field of education including parents, students, and key decision makers and politicians, the battle for a person's mind to consider technology as a key, important subject within our schools is lost before it has even begun. This realisation pertaining to the importance of advancing technology teaching becomes a positioning problem because it is an initiative much bigger than any one person can advance. These, and other observations, are the very reason that associations exist today, and will in the future.

Associations exist in many different forms around the world for the advancement of technology education. They exist in all levels of maturity, from very beginning associations to those that have made significant contributions to our profession. Each has its own culture, or personality, as a result of inside and outside influences affecting the level of significance of technology and the way it is taught in schools. Often, technology taught in a specific country is reflective of the level of technological sophistication being used in that country.

Similarly, associations tend to reflect the philosophies for delivery of instruction that dominate in particular educational jurisdictions. For example, the United Kingdom associations have a primary interest in advancing "design and technology" while "technology and engineering" is a major emphasis for United States associations. In Canada and Australia, provincial, territorial, and state associations seem to be more active and stronger than national associations. Associations in developing countries, if in existence, have a tendency to evolve as their education infrastructure grows.

The strength of associations grows and wanes depending on positive and negative controversy related to action both inside and outside of the profession. Association leadership processes create positive controversy through attracting members, engaging individuals, soliciting information, and attaining general consensus on curricula initiatives. These leadership processes create communities of practice, core values, ways of operation, and methods of engaging teachers, administrators, and other key stakeholders. Standards are built by using focus groups to either develop the initial frameworks or seek reaction and improvement by stakeholders or teachers who are responsible for delivering instruction. New curriculum directions are identified or refined through similar processes that create positive controversy. The end result is hopefully a document that reflects the best of collective thought. In other words, technology education associations have become leaders for their teachers by creating documents that establish a direction or territory of the curriculum that belongs to their field. This is important, since teachers are key to advancing the profession. If they have not bought into ideas and next directions, any proposed progress will fail.

On the other hand, failure to change names or curricula to keep pace with broader developments, lack of success in getting included in progressive legislative reform, or being included as a sub-part of another subject's new standards are examples than can cause negative controversy, hindering the growth of technology education. If an association cannot move quickly to turn a negative ramification into a positive direction, membership interest will drop and, where more than one association exists, members may even migrate to a competitor association that is perceived to better represent the needs and concerns of educators.

A recent example of potential negative controversy may have happened in the United States, with the science community creating standards for technology and engineering education under the umbrella of science education. If this standards thrust is successful, it will cause the national science teachers association to incorporate the technology and engineering community into their organisation. At the same time, the technology and engineering associations will have to make adjustments if they are to hold their leadership position with their own teachers and subject area. It will take years to judge if this move by the science community was fruitful for them or their proposed partners from the technology and engineering community. Hopefully, it will be fruitful for both association communities.

The political structure within a given country also affects the type of association that may exist because of the way that decisions are made at the various levels of government. A democratic form of government will result in an association operating in a different way than associations in countries with authoritarian governments. Nevertheless, many of the same problems will exist when attempting to teach technology in schools, for example, teacher preparation, instructional strategies, assessments, and more. There is, therefore, a common bond between technology teaching professionals regardless of a country's culture or political structure. Within this, associations can become the common ground for the exchange of ideas and initiatives. Often what works in one country (e.g., using design as a key curricular component) is tried in others, some of which have entirely different cultures or political systems.

## The Power of a Name

Successful association boards in a democratic society have two main purposes: to determine directions and to set policy. Successful educational associations therefore have a positive mission and purpose to further the ideals of the profession. They aim to help their members grow professionally and to advance thought and practice pertaining to high quality technology teaching. This effort is a way of moving the profession from its 'current reality' to one that is more visionary.

One example of an association 'determining directions' and later 'setting policy' happened with the American Industrial Arts Association (AIAA) in the United

States during the 1980s. The association leadership needed to anticipate a new curriculum for a society that was moving from being industrialised to one that was becoming more technological in nature. Thus, an association name change was proposed after membership surveys, forums, and discussion amongst the Board of Directors. The change from Industrial Arts to Technology Education caused considerable stress because of comfort with the decades-old name, Industrial Arts, which was well known within educational circles.

The AIAA Board of Directors knew that they had to be careful about perceptions, for the selected name of the association would also be the title of a new curriculum. This new curriculum would be developed and promoted both in and outside educational circles, probably for decades. Such a name change would therefore have to be accomplished with thoughtfulness and concern, taking into account the smallest of details. For example, a name such as 'Industrial Technology Education' at the national level could result in a funny or negative acronym with affiliates at the state or local level, causing the subject name not to be readily acceptable.

Ultimately, the name, International Technology Education Association (ITEA), was chosen because it best signaled the change towards the new technology education curriculum. Members had become aware of the presence of 'technology education' as a term in the literature and the name of the association's journal had recently been changed to *The Technology Teacher*. ITEA more recently made a second name change to include engineering in their association's title: International Technology and Engineering Educators Association (ITEEA). This change reflects political positioning to include engineering education at the K-12 levels of education, while the previous name change was more to signal a change in curricular direction. The second name change better positioned technology education to be a player in the technology and engineering standards that were later created by the science community.

## School Politics

Teachers place little emphasis on school politics at the primary or secondary level where the real action of teaching and learning is focused. Politics at this level tends to be focused on teacher unions or bargaining units addressing salaries and other benefits that accompany teaching in school systems. Of course, this is a primary factor in any teacher's life because of financial and health impacts on a teacher and the teacher's family. This chapter will not address salaries and benefit negotiations, but will address politics and policy making tied to technology education.

Teacher preparation usually does not educate teachers about the attributes of associations or the dynamics of being politically active to support technology education in schools.

School teachers normally have to learn about politics in a random fashion, usually from colleagues with the same teaching and content interests. For example, technology teachers are often interested in technologies such as robotics or solar energy. They know that they can gain knowledge in these areas from meetings and training. The meetings are usually organised by their colleagues who are members of a local, regional, state, provincial, or national association. The need to know more about teaching and the technical expertise to be able to teach about design, technology, and technological literacy concepts causes the teacher to take initial steps towards becoming involved in association activities.

As school systems change curriculum, standards, instructional media, assessments, and other related items, teachers are expected to become knowledgeable in their use. For example, school systems often have some type of learning standards or the need for assessments written in a specific manner related to technology education. Associations often have anticipated such needs and have models to share. The degree of passion that a teacher has for staying current with the latest educational developments is often directly related to their amount and degree of association involvement. Obviously, educators active in an association are better prepared to be excellent teachers simply because they have been involved in the development of materials and strategies used by the association and technology teachers. These involved educators also become recognised as leaders because of the knowledge they have gained through association participation.

It is clear that a group of educators often have more influence on an issue, problem, or opportunity than a single person. Therefore, a group of technology teachers can have more influence on key decision makers than a person representing a single programme. Associations often use this numbers power to influence educational leaders beyond their subject area, for example, in calling for lower student:teacher ratios, organisation of technology teaching within a STEM unit of a school, changing the philosophical direction of an existing programme, calling for more professional development support, stopping spending on outdated initiatives, etc. Associations become a way of organising the masses for the good of the majority involved. This is particularly true when asking school boards, government leaders, or elected representatives for more funding to expand the impact of technology education.

The needs of the technology teacher become the major focal point for any effective technology education association. The support and time given by the classroom teacher to an association usually results in knowledge gain by the teacher, networking advantages, and political support as a result of being known and interacting with other leaders. A political support system can be the difference between a programme surviving during tough financial times or being terminated as a result of no one being politically savvy enough to stop the termination process.

The ideal politics and policy-making programme of an association at any level is one that promotes positive ideas that will help educators and their students. Such a programme would undertake offensive initiatives for the growth of technology education, rather than constantly being on the defensive and trying to save past initiatives. School politics is one of the most important areas of technology education because it can do more to affect the overall health of technology education programmes, and it all starts with the actions of teachers.

## Worldwide Networking

Worldwide social networks also have a tremendous influence on perspectives and decision making simply because information is shared so quickly and can influence each person. This type of networking often forms or is directed through an association. Professionals are also often directly connected to the association(s) that they value from around the world. For example, it is not uncommon for a professional from Africa to be involved with associations and meetings in Europe, or New Zealand educators being active in Japan, China, or Finland.

Technology education researchers from around the world share philosophical positions and research through journals and selected other publications, and interact during research conferences, such as the Pupils Attitudes Toward Technology (PATT) conferences or the Technology Education Research Conference (TERC), where ideas, directions, and philosophies are tested. For example, the many characteristics of technology and engineering as it relates to science and mathematics will be shaped at conferences related to science, technology, engineering, and mathematics (STEM) education. These face-to-face meetings are further supported by electronic networking, which advances discussion in terms of practice and thought, or simply by watching what fellow professionals are doing.

Of course, social networking has also forever changed the face and internal operations of associations (Nour 2011; Sladek 2011). Although we may not know what social networking will look like in the future, its effect is going to be huge. For instance, the idea of "belonging" to an association is changing in that members will belong in the future, but in a different way. Perhaps accessing resources from associations will become like a person walking up to a vending machine where money is paid for the product and the person walks away with the product. This is an important adjustment for associations because they may need to change how their primary revenue is derived.

Within such a changed context, it may not be how many members an association has that is important, but rather how much influence an association can create for their cause. Yes, membership numbers count, but the old days of representation according to a membership infrastructure is fast becoming passé. Professionals are no longer joining associations for the same reasons as in the past. Few care about the degree of representation that an association can deliver. Many do not care if they become an association member. They want what will help them in their current situation and they want it NOW. If they cannot get their wants and needs satisfied by the association, they will go to other places on the internet. This places technology education associations in a precarious situation. They still need members to provide the core association direction and financial income, but their professional members do not necessarily have the same desire and commitment as in the past to affiliate.

## Association Culture, Core Capabilities and Forward Thinking

If associations are to act as a major player in the politics and policy advancing technology education, the way that technology educators think about their future professional association will have to change. This change must happen because of progress that will be made in communities of practice with the next generation of teachers' core values, ways of operation, and ways of engaging others. The use of information technology, often by younger professional members, has already changed that thinking. The older members must move quickly to obtain a new mentality about technology and its use in communicating if the association is to attract younger members. We know that electronic communication will continue to change technology advances, and that it changes the way people communication and think about society.

Changing the association culture is not an easy task, since most associations have a culture that is adverse to change (Coerver and Byers 2011). Volunteers, such as board or committee members, don't want to disagree with peers and staff don't want to disagree with the leadership, who are the volunteers. Technology educator associations often focus the majority of their time on procedural or managerial functions, such as keeping the committees, task forces, and Boards functioning. Often so much time is spent on operational functions that little emphasis is placed on where the time and energy investment should be placed.

Moving forward, associations will likely need to drop most of their procedural and managerial activities and concentrate only on what they do best. This can be done by first asking and answering where the association can best invest its time and energy. Core capabilities must be identified and only a leadership commitment to get maximum results will create a culture effecting forward thinking. When technology education association cultures fail to change with the times, the politics and policies needed to advance technology teaching will be severely limited or may not occur at all. The result will be many missed opportunities.

What are association core capabilities? Often, only the association members have the knowledge and experience to determine core changes related to their region of the world. At selected times, a core capability will arise and be so obvious that it becomes evident to all and automatically attracts support. Such an example is one that evolved in the United States pertaining to drunk driving. So many individuals were causing accidents and deaths as a result of drunk driving that it caused the creation of an association called Mothers Against Drunk Driving (MADD, www.MADD.org). MADD worked to have state and national legislation passed to create stiffer penalties. They created greater awareness about the drunk driving problem and contributed to bringing this more under control for the betterment of society. MADD serves as a model for identifying a core issue with the ability of creating an association to do something about it. MADD coalesced around a leadership group who tackled the problem, and attracted and mobilised many others to help work on the solution. Although the reason for the creation of MADD has been addressed, the association continues to keep the problem of drunk driving in the forefront of North American society. Technology education associations need a similarly compelling core reason (s) that will attract educators from around the world to an equally important mission. Such a compelling issue has been hard to identify.

What are select core capabilities for technology education associations and professionals? The answer may be found in the reason that technology education associations were created in the first place: to advance thought and practice about the teaching of technology and technological literacy. However, this reason alone is not sufficiently compelling to cause masses to be mobilised into action. A problem or issue related to technology and technological literacy that is more compelling needs to be identified. For example, if there was a direct link drawn between technological knowledge and/or capabilities and how much a population thrives as a nation in terms of goods and gross national product, more interest may be focused on teaching technology. The end result could be better positioning and more support for teaching about technology, innovation, design, and engineering. However, even this direction may not be compelling enough, for mathematics and science are often given credit for such advances, not technology.

The identity, positioning, and advocacy issues associated with the technology teaching profession remains a hindrance to the growth and support of technology education. Unlike the school subjects of history, science, or mathematics, where much of the content is based on constant facts or theorems, technology is dynamic in nature, with changing, new content that reflects the latest developments in society. The profession's content once covered woodworking and metalworking, but now covers such topics as computer-aided design, robotics and lasers. These changes make it difficult to keep the various publics apprised of the latest technology education identity. In other words, identity and positioning problems are likely to continue within the larger community of education, as well as with parents and other stakeholders. Technology education associations inherit these identity, positioning, and advocacy problems.

At the same time, the compelling reason for starting an association may not remain as compelling for keeping the association moving forward in the long run. Other more compelling core capabilities need to be constantly identified, reviewed, researched, trial tested, and implemented for a profession and its representative associations. These characteristics help an association to remain vibrant. The longevity of technology education associations will depend on the compelling nature of their core capability. As previously stated, this is not a situation that is unique to technology education associations. However, it is a big challenge for the profession and its associations if political action and advocacy efforts are to be effective change agents.

Perhaps one mandate is for technology education associations to have a mission of developing essential resources for improving teacher performance, while at the same time creating an environment that will allow the association to stay in business. This culture should look to and reflect on how an association adds value to the work of technology teachers who are the membership—if membership is still composed of only technology, design, or engineering teachers. It will likely be necessary for the association leaders to strive for deeper, more meaningful services that might include project information or selected services that matter to members, add depth to current services, or create new services to address new member needs. Some technology education associations might be in the position to do less, but do it better.

Technology education associations must address the characteristics identified in the previous paragraph as signs of change and look at the traditional membership association as the end of membership as we have known it (Sladek 2011). Today's association culture signals, with its many electronic capabilities and internet delivery of services, a dominant, compelling purpose to attract a new membership that will be teaching in a similarly electronic environment. At the same time, it must be a revenue generation association that stands out with a uniqueness that makes it different from other subject area associations. The future association will have a culture made up contemporary values, beliefs, assumptions, experiences, habits, and robust mechanisms for supporting members working together in an electronic world. These characteristics will likely all be required in the race for relevance for technology education associations.

## Raising the Value of Technology Education

Earlier in this chapter and in other chapters of this book, the need for quality assessments and indicators have been noted as critical components for the future of technology education (see Chap. 7). Such indicators provide fuel for political presentations in which it is important to cite data proving the worth of technology education. These 'quality indicators' can provide selected 'vital signs' to change technology education in both politics and policy making, showing a subject area that provides many positive attributes in the education of all students.

Technology education associations have played significant roles in helping the profession obtain research funding, articulating research findings both inside and outside of the profession, advancing ideas and practices, promoting researchers, universities, and their projects, and conducting research using association staff. However, the task of making technology education of value in politics and to policy makers can more easily be expressed than accomplished. Enough people need to be convinced that a strong knowledge of technology is an important requirement for their daily lives, that they cannot live without technology, that it is imperative to educate a next generation of technologists, innovators, designers, and engineers who will become the thinkers and leaders of tomorrow-and, that such an education comes primarily through technology education.

This task needs to be addressed with all of the enthusiasm and energy that is used to elect a nation's president, prime minister or premier. It involves all of the work of a master political strategist and a constant, unrelenting quest to provide the best technology education possible to students currently in school. Work is needed to make technology teaching a 'personal issue' in the lives of everyone, including politicians and policy makers. The task at hand is monumental. Making technology education of such value would have significant positive impacts for any country in terms of thinkers, makers, inventors, product designers, and innovators leading in a successful thriving environment. The country that creates the next generation student with these qualities will be well positioned to become a world economic, political, and societal leader.

How does the field of technology education reach such a momentous achievement? The answer relates to the ability of the field to be distinctive in the school curriculum and in creating a positive perception in the minds of parents and decision makers. It is, in other words, an issue of branding (Starkweather 2011). In addition, discords between the aspirations for the subject and its actual delivery in the classroom must be narrowed. The public needs to be educated about what technology education is. Technology educators must help people trust and believe in the worth of technology education, creating the desire and need for the subject in schools. Technology education does not have a bad image; it has little or no image.

Technology education must look closely at the essential characteristics that describe the subject and then start shaping a culture of what they want to be. The idea of the subject as being all things to all people must be addressed by a specific focus for the subject. Patience will be important while striving for consistency, staying focused, believing in ourselves, and providing strong models based on research. The charge is one of developing a profound sense of mission and sharing it with anyone who is willing to listen, and in some cases with those who aren't.

Who, then, are the stakeholders of technology education? The short answer is "everyone". However, the real stakeholder may be "you". Any person can help empower the teacher, who in turn empowers the student through learning technology. That student then becomes the future engineer, teacher, architect, skilled tradesperson, and other types of technological workers. The distinctive characteristics of technology education addresses our ability to tinker, design, create, critique, make, invent, and attempt to better ourselves and the environment, affecting our culture and world. The human being is constantly trying to satisfy wants and needs by creating, shaping, and adjusting technological worlds.

But not all students will be employed in a technology-related field. Many will carry their technology learning with them into their futures. They will improve their technological capability, design and invent, become innovative in many ways, and address societal problems through technological solutions.

Technology educators want their subject to be a valued part of a student's overall education. They want to provide education in a quality way by creating the next generation of technological thinkers. Achieving the desired perception requires hard work and dedication to ideals, being advocates of what we do, and staying informed about the latest education and technological developments. Attaining these public and professional characteristics requires more than an 'average' advocate for technology education. It requires a large group of advocates who are unrelenting in their desire to have a valued, strongly positioned school subject. That large group of advocates is very often a technology education professional association.

Often the potential allies for technology education do not know that the subject exists. This is a major disconnect. For example, corporations who could benefit from having employees with a technology education background are often totally unaware of this type of school education. Rather, it is assumed that a student who has taken mathematics and science also has a fluency in technology and engineering. The field of engineering often perpetuates this misnomer by requiring that students take all of the mathematics and science courses that they can take in order to become an engineer. While this is important, technology education may be the main reason that a student chooses to become an engineer or technologist in the first place. The technology education profession and its associations have a major ongoing task of creating allies with corporations that desire this type of learner.

Selected charitable foundations have provided their support for technology education through funding, such as the Gatsby and Nuffield Foundations and Wellcome Trust in England or the Foundation for Technology and Engineering Education and the Technical Foundation of America in the United States. All of these support leadership initiatives to advance technology education and have done much to help the profession in their respective countries.

Other allies of technology education, who also are often overlooked, include the national academies of technology, engineering or science, which commission various studies to address opportunities and initiatives that will support and further define technology education. These academies often work with associations, foundations, and government ministries to advance STEM education.

## Showcasing Success

Nothing creates value like success. Student achievements will tell the best stories for technology education. Such success should not only come through competitions, but through measurements that the public and educational systems value. These successes should be articulated with a rigorous publicity campaign that goes beyond the school walls. Often, it is the association that articulates these successes throughout the educational community and to the general public.

Positive and trusted images of technology education must come from what we do through student achievement, which should be showcased at every opportunity. In a related vein, systematic, sound research about teaching and learning should inform teacher developments, further strengthening their ability to adjust and reform technology education. Constantly striving for such successes and informing others of the quest for quality technology education will eventually lead to a valued subject area that strongly contributes to the education of all students.

Making technology education of value begins with passionate leaders who start small, but have a vision, passion, and the leadership ability to build groups focused enough to create success beyond what could be imagined. Corporate leaders such as Ray Kroc of McDonalds, Richard Branson of Virgin Atlantic, and Sam Walton of Wal-Mart are excellent examples of such leaders. They were builders of something that became larger than they probably ever dreamed.

Every teacher can become a passionate leader of technology education, building something larger than they can imagine. Causing technology education to become an integral part of every person's basic education is the life's work of a technology teacher. As with other passionate leaders, technology teachers need to lead the cause in making technology education a valued education that is desired and sought by parents and policy makers. When the politicians and policy makers are upset at potential cuts to technology education, members of the field will know that they have been successful at making technology education of value.

## Summary

This chapter began with a quote from Einstein and the idea that success breeds value. To be valued is to be meaningful, relevant, beneficial, significant, and a priority. Being "valued" is one of the major challenges that technology educators have faced during the relatively short history of their profession. Associations have played a key role in way that technology education has advanced over the years to where it is today. Associations have created an opportunity for technology educators to work together to address issues and advance causes strengthening their profession.

Technology education advocates have had a history of positioning and repositioning technology education. They have changed the names of their profession as technology has advanced. They have increased funding, advanced educational initiatives, strengthened data gathering, provided research to guide the profession, and worked to raise the value of technology education as a part of the overall education of a student. These successes are noteworthy and must be continued if the profession is to be vibrant in the future.

Future technology education associations will have to make the teaching of technology a personal or core issue and create an association with a sense of purpose that consumes one's professional career if the association is to be productive and viable. Association leaders will have to constantly search for what that personal or core issue(s) is to remain a productive association. The issue(s) may vary from country to country, with the common denominator of needing a compelling issue causing one to join and be involved. It is the heart of future association membership. The core issue is the 'life's work' of an association, driving politics and policy making to advance the profession.

A tremendous amount of work has been accomplished to progress technology education to this point. Now, with the changing nature of society, technology educators must adapt and capitalise on the qualities produced by this type of education.

The opportunities to influence the politics and policy for technology education are vast. The amount of work to be accomplished is great. The stakes for technology educators—and ultimately students as future workers and citizens—are high. The reason for what must be done is very clear. No other field can show how technology and technological literacy can be taught to prepare young minds toward a creative and satisfying life like technology education.

## References

Coerver, H., & Byers, M. (2011). *Race for relevance: 5 radical changes for associations.* Washington, DC: American Society for Association Executives.

Nour, D. (2011). *Return on impact: Leadership strategies for the age of connected relationships.* Washington, DC: American Society for Association Executives.

Sladek, S. (2011). *The end of membership as we know it.* Washington, DC: American Society for Association Executives.

Starkweather, K. N. (2011). Branding: Putting a little dent in the universe. *Technology and Engineering Teacher, 70*(6), 36–40.

# Chapter 14
# Research Challenges for the Future

**Marc J. de Vries**

Starting with a brief review of research in technology education, this chapter goes on to propose research that continues to be needed in the context of technology education, a school subject that continues to have uncertain status and a problematic image. While remarkable progress has been made in technology education research over a relatively short period of time, significant work remains. First, research questions and research findings need to connect more closely with teachers. An important possibility here is involving teachers more closely in the research. Second, targeted policy-oriented research is needed and policy makers need to be recognised as an important audience for future technology education research. Third, more sophisticated research is needed on how to better support students' technology learning. For this, a design-based methodology may be particularly fruitful. The extent to which researchers are able to realise closer links between their work and educational practice, and enhance their understanding of policy processes, will likely significantly impact the future of technology education.

## Introduction

In this chapter I will present a perspective on future directions for technology education research. In doing so, I will first look back on the (short) history of technology education research and show where this has brought us. In the course of time, it has become evident that technology education research still needs some careful rethinking to be really effective in the context of a still developing school subject with uncertain status and a problematic image in many places. Although

M.J. de Vries (✉)
Delft University of Technology, Delft, Netherlands

Eindhoven University of Technology, Eindhoven, Netherlands
e-mail: m.j.devries@tudelft.nl



PJ. Williams et al. (eds.), *The Future of Technology Education*, Contemporary Issues in Technology Education, DOI 10.1007/978-981-287-170-1_14

technology education research cannot on its own address all the problems, there is certainly potential for contributing—but in order to realise this, new directions will need to be taken.

In this chapter, I will pay special attention to the issue of the research-practice link. This link is problematic not only for technology education, but for other school subjects as well. However, the often fragile position of technology education in the curriculum increases the need for this link to work well. This is related in part to the contribution of technology education research to the justification for having technology education in schools, a point also raised in Chap. 13 by Kendall Starkweather. While it may seem that this motive for doing research is fairly defensive, the current situation in technology education justifies such an approach. It does, however, not mean that there is not also an important 'offensive' role for technology education research, namely to support the further development of technology education by providing insights into what works and what does not work in teaching and learning about technology. This is probably what most researchers are themselves primarily interested in, and justly so. Further, an enthusiastic researcher who comes up with stimulating ideas for improving technology education practice is also the best defence agent for the school subject. In designing and organising technology education research, a focus on directions that will also enhance the defensive role of the research is timely.

## The Road Travelled So Far

Research in technology education is a relative newcomer in the educational research domain. This is due to the fact that technology education is relatively new in the school curriculum. While there have been craft-type school subjects for a much longer time, technology education is generally understood to be far broader. Technology education entails not only training in manual capabilities, but also design capabilities, knowledge development and attitude formation—and it is the design and knowledge components in particular that gave rise to an interest in investigating technology education. This is somewhat surprising since even a subject that only aims for manual skill development offers good reasons for investigating how this can best be learned and what pedagogy is needed. For some reason, and probably this is simply the undervaluing of manual skills in general, academic researchers have seldom taken an interest in this. However, as the design and knowledge component have increased in importance, research in technology education has emerged as a separate research domain. This has happened in a fairly short period of time—only three to four decades. Before then, there were no academic research journals specifically for technology education; neither were there international conferences dedicated only to technology education research.

In the late 1990s, some review studies were undertaken to determine the kinds of research studies that had been done up until then (de Vries 2003; Foster 1992; Petrina 1998; Zuga 1997). The results were fairly disappointing: many studies were theoretical, focusing on the identity of the subject (understandably, given its newness), curriculum content and educational goals for technology education. In terms of methodologies, the range was small. Most studies at that time were quantitative in nature. Studies into classroom practice were few. Most research was done in the UK and USA. While there has been no systematic investigation of changes since then, my impression as the editor-in-chief of the *International Journal for Technology and Design Education* is that the number of theoretical studies has dropped significantly (to almost zero) and the number of qualitative studies in classroom contexts has grown from almost none to a fairly steady flow of articles.

In terms of topics covered in research, three main areas can be distinguished (de Vries 2003): (1) research into what is to be taught (standards, curricula), (2) to whom and by whom it is taught (pupils', students' and teachers' attitudes, knowledge, skills, social background, etc.), and (3) in what ways it is taught (pedagogy, use of media). In the early years there was a lot of attention on the content of the curriculum. For example, there was interest in comparative studies in which approaches in different countries were analysed. Gradually an interest in learning from the philosophy of technology, design methodology, and from the history and sociology of technology emerged. Some, like Sven Ove Hansson, John Dakers and myself, even became active in both fields to enable immediate transfer from such domains to technology education research.

Later, studies into the practice of conducting design projects in technology education appeared in research journals. It was particularly in the UK that such studies were carried out, not surprisingly because design was the heart of the school subject Design and Technology. In the context of research into design projects in schools, various sub-topics have been covered, such as drawing (2D and 3D, by hand and using CAD programs). One topic in particular that has received considerable attention is the assessment of design projects. Here, the efforts of the research team at Goldsmiths College Technology Education Research Unit, led by Richard Kimbell, should be mentioned. This research line continues today and has led to numerous useful insights (see, for example, Kimbell and Stables 2007 and Chap. 7 of this volume by Kay Stables). Independent of all this, research into design education has slowly emerged. In leading design methodology journals, such as *Design Studies*, articles on design education research are published. Such articles can also be occasionally found in journals for engineering education. In the future, design education research may become another relevant field to learn from, as has happened with the philosophy of technology and design methodology.

The current situation for technology education is that we have several academic journals: the *International Journal for Technology and Design Education* (published by Springer, a commercial publisher), the *Journal of Technology* (published by the International Technology and Engineering Education Association in the USA), the *Design and Technology Education: An International Journal* (published by the Design And Technology Association in the UK), *Studies in Technology Education* (published by Epsilon Pi Tau, a North American fraternity) and recently the *Australasian Journal of Technology Education* (supported by New Zealand and Australian technology teachers' associations). In addition, there is a series of conferences in which research in technology education plays an important part (such as the Pupils' Attitude Towards Technology Education—PATT—conferences, the DATA Annual conferences in the UK, the TERC conferences in Australia, and the PATT sessions at the Annual ITEEA conferences). Taken together, these outlets show that research in technology education has matured to a certain level. This was a reason to set out the status of the domain in the *International Handbook of Research and Development in Technology Education* (Jones and De Vries 2009).

Although the remainder of this chapter will describe ways in which technology education research will have to take new steps moving forward, I want to emphasise here that what has grown in a relative short period of time is remarkable. Research in technology education is a well-established discipline with a high quality that meets scholarly standards.

## The Research-Practice Link

One of the urgent challenges for future research in technology education is to create a better connection with teaching practice. This is not only a challenge for technology education research. For other educational research, too, a complaint can often be heard that it is too heavily oriented towards developing theoretical insights, interesting though they may be, while failing to have an impact on educational practice (Broekkamp and Van Hout-Wolters 2007; Wicklein and Hill 1996). This can arise for several reasons. In the first place, the research questions are often developed by researchers and teachers have almost no say. They are merely the people that provide research data. Additionally, in the analysis phase, it is the researchers who tend to make all the decisions. This is understandable, because the researchers almost by definition are the ones who have the knowledge and expertise to make these decisions. But the consequence is that opportunities are missed to derive research questions directly from teachers' experiences: What do they see as problematic? In the publication phase, the researcher's priority is to get the material published in an academic journal, as this is what they have to account for professionally: their number of international scholarly articles. Writing for a teachers' magazine is viewed as 'wasted time' by some researchers, even when they are aware of the need to inform teachers about the outcomes of the research. Teachers do not read academic journals, as a rule. In addition, it can be difficult for them to understand the content of what the researcher has written, and even more difficult to see what the research might mean for their practice.

It would be an interesting exercise—but beyond the limits of what I can do in this chapter—to compare what is in teachers' journals with what is in research journals in technology education. Being a reader of both, my estimation is that a striking mismatch would be found. Teachers' journals (for instance, the ITEEA's *The Technology and Engineering Teacher*) are full of ideas for new topics in and suggestions about how to teach these. Such ideas are largely absent in research journals. In addition, not many scholarly articles use a design-based research approach in which a concrete intervention is designed, its effects investigated and the outcomes used to improve the intervention, which would give a better connection to what teachers are interested in. I will come back to this approach later.

There are several possible responses to the issue of the gap between research and practice, and the minimal impact that research has tended to have on practice. In the first place, research could be better embedded in the whole process of educational development. Organisationally, it is often isolated from other parts of the development chain. At best, a research programme is attached to a teacher education programme, and even then the researchers are not necessarily the same as those who are involved in teacher education. It would probably require additional education for both researchers and teacher educators to be able to exchange places in a 'research and teacher education programme', but it is certainly worth seeking closer relations between the two activities.

A second response to the issue of low impact on practice is to more closely involve teachers in research. This can be done in the traditional types of exploratory and experimental research. Teachers could actively be involved in the processes of problem definition, selection of research methods, collection and analysis of data, and concluding the outcomes. Of course their role cannot be to replace the researcher's methodological expertise, but they can provide a meaningful addition by bringing in their experiences from practice. Scientific research focuses on specific aspects of reality and that concentration makes it strong in that it allows in-depth investigation. Life, however, is a complex mixture of multiple aspects and classroom teachers tend to have a stronger awareness of this knowhow than researchers. They can help researchers identify 'blind spots' that result from their attitudes as researchers to confine their study to certain aspects only. This does not mean that researchers have to give up their focus, but it can help to get the focus 'right' by choosing from a broader range of possible aspects.

The most intense mechanism for involving teachers in research is through educating them to become researchers. Several countries, such as Sweden and the Netherlands, offer teachers an opportunity to do a Ph.D. funded by government. In Sweden this has led to two separate research 'schools' for (science and) technology education, FontD (Forskarskolan i naturvetenskapernas, teknikens och matematikens didaktik; transl. National Graduate School in Science, Mathematics and Technology Education Research) and TUFF (Teknikutbidning för framtiden; transl. Technology Education for the Future) (Skogh and Gumaelius 2012). These have resulted in a series of Ph.D. dissertations. After finishing their research study, the teachers return to schools with new capabilities that can make them better teachers in their own schools, and better equipped for working with university researchers in the future. Also in some countries, like in the Netherlands, there are 'academic' primary and secondary schools in which teachers do research based on a school-wide research plan that is defined by the school. Early experiences with this show that the quality of the research is often poor and the activity should be regarded more as a form of professional development for the teacher rather than a serious academic research effort. Even so, it can still work as a mechanism that improves the relation between academic research and teaching practiced in that teachers have a better understanding of researchers' work.

A third response to the problem of the missing link between research and practice goes even further and deals with the very methods that are used by researchers. All previous suggestions are still limited compared to what is probably the most fundamental approach to the problem, which is to shift to a whole new research paradigm, of which design-based research is the best candidate (see The Design-Based Research Collective 2003). Design-based research should almost by definition appeal to technology education researchers, as design is a vital element in technology education.

Design-based research means working as a designer would do. The purpose of design is twofold. In the first place, a designer aims to bring forth a new artefact, system or process. In order to reach that goal, a prototype is made, tested, and the outcomes are used to improve the prototype, which then is tested again until a satisfactory design is reached. But apart from the artefact, there is a second purpose fulfilled, namely that the designer has gained new insights that are tested empirically. During the experimentation on the prototype insights into the relations between properties of the prototype and the behaviour of the prototype are gained. This knowledge is transferrable to similar situations. In the beginning this knowledge will be fairly 'local', that is, specific to situations that are very similar to the original prototype that was tested. But by building up a series of design experiences, larger numbers of opportunities to generalise this 'local' knowledge emerge and more generic knowledge is developed. This is also very much the way knowledge is developed in engineering sciences. The methodological problem is that it is not easy to compare the prototype to a 'control' situation, as too many variables change at the same time, which is deadly for a true experimental or quasi-experimental set-up. One can similarly question how problematic multiple variables are for educational research, where very large numbers of classes and pupils are needed for experimental or quasi-experimental research to compensate for all the background noise caused by the fact that classes differ, teachers differ, circumstances differ, etc. If one critically examined current educational experimental and quasi-experimental research, one would, no doubt, find many cases in which the reliability and validity of these studies can be questioned on the basis of this consideration. This indicates that the loss of precision due to a shift from more traditional and quantitative research methods towards more design-based and qualitative set-ups may be much less than one would estimate at first sight.

The value of design-based educational research for practice is, however, much higher than for the more 'classic' studies. Here, too, there is a twofold benefit: an improved educational outcome (a new teaching pedagogy, new lesson material, new media, etc.), that is, the design. Second, there is the additional knowledge that was gained during the testing of the 'prototype'. While it may seem more difficult to get these sorts of studies published, we can already see more journals accepting such studies. This is partially because qualitative research, in spite of its original challenges in identifying validity and reliability, is now generally accepted as a legitimate form of research. The activities of the Design-Based Research Collective (www.designbasedresearch.org) have been helpful here, their studies showing that this type of research can have high validity and reliability, albeit in different ways to more traditional experimental or quasi-experimental quantitative studies. In addition, action research, which is very similar in nature to design-based research, is gaining ground. The advantage of design-based research over action research is that the former necessarily results in concrete and usable outcomes for educational practice, whereas action research may still remain in the realm of theoretical interests. Design-based research is, therefore, a very attractive option for investigating how design and technology activities can be optimised in their learning effect, particularly where both skill and conceptual development are targeted. In this case, design is both in the method and the content of the research.

The methodological debates referred to above reflect a tension between two perspectives on the relation between research and reality. Traditionally, a realist stance was taken and research outcomes were assumed to be a one-to-one mirror image of reality. More recently, an alternative approach has emerged in which research outcomes are understood to reflect more of the researcher's perspective of reality rather than an objective reality independent of the researcher. The view I take in this chapter is that both are invalid. My stance is that of 'soft realism', in which there is awareness that there is a distance between research outcomes and reality, caused by the interpretation that is involved in data collection and analysis. But soft realism believes that this distance does not mean giving up on the possibility of gaining knowledge about a reality that exists independent from the observer. I think soft realism is in a better position to justify educational interventions based on research outcomes. Finally, it is useful to point out that the dichotomy should not be confused with the difference between quantitative and qualitative research, as if quantitative research is necessarily connected to a naïve or soft realist stance and qualitative to a postmodern constructivist stance. Both types of research can be used in both stances.

## Policy-Oriented Research

As pointed out by Kendall Starkweather in Chap. 13 of this volume, technology education research can play an important role in dealing with one of the subject's major struggles, namely evidencing its impact on pupils and society. Policy makers are often impatient and want to see concrete evidence that technology education at least to some extent fulfils its intentions. In the past, we have been quick to claim that technology education develops technological literacy, creativity, communication and cooperation skills, and that it supports science, mathematics and language learning. But does all this really happen in practice? We still believe it to be true, at least to some extent, but the evidence is not so definitive.

It seems that it should not be too difficult to collect evidence of the impacts of technology education, as the research methodology can be fairly straightforward. Large-scale quantitative studies should be able to do the job, as has been tried in the past with quantitative pre- and post-test set-ups. In the report Tech Tally (Garmire and Pearson 2006) an extensive survey of existing instruments for assessing technological literacy was presented. Importantly, it showed that either the instruments assess higher order skills and are too complicated for large-scale quantitative studies, or they are suitable for large-scale studies but do not assess higher order skills. This does not mean that large-scale assessments of higher order skills are impossible, but it does show that efforts to assess higher order technological literacy skills—an important desired outcome of technology education—have thus far been unsuccessful.

It has been difficult to develop valid instruments for testing high-level capabilities, such as means-ends reasoning, cause-effect reasoning, decision skills for responsible citizenship, etc. Therefore, while technology is now included in TIMSS, there is concern that this way of testing outcomes of technology education is far too simple to reveal what really has been learned. It is, for instance, virtually impossible to assess design creativity skills by means of a standardised paper-and-pencil test. This type of skill, however, is very important in technology education. The work of the Assessment Performance Unit at Goldsmiths in the UK has shown that assessing design skills require more sophisticated instruments, such as portfolios (Kimbell and Stables 2007, see also Chap. 7, this volume). Nevertheless, policy makers will likely look at the outcomes of international studies like TIMSS and draw conclusions with respect to the future presence of technology education in the curriculum.

In this respect, studies such as PISA and TIMSS seem to do more harm than good for technology education. For example, studies in the context of the ROSE project (Sjøberg and Schreiner 2010) have shown that they should be read with care as there are quite concerning findings, such as the inverted relation between scores of PISA and interest in the subject: often the better scoring pupils do not like the subject and for this reason may drop it as soon as they can. This, then, highlights the ongoing need for the development of valid and reliable instruments for measuring higher order technological literacy dimensions (knowledge, skills). It does not mean that TIMSS and PISA studies can be ignored in the field of technology education, as they do have a certain status. We should, however, not focus on getting high TIMSS and PISA scores as an aim in itself, but rather hope and expect that good technology education will also result in positive outcomes in studies that have been more broadly conceived.

Policy makers should be recognised as an important audience for future technology education research, but more will be needed in order to get their continuous support for strong positioning of technology education in the curriculum. What also needs to be done is to develop another new type of research study—one that focuses on the process of educational change, with particular attention to the role of policy making. In general education research, this is not new. Such studies have been reported in the past and they are still being conducted. However, they do not give specific insights into how this works for technology education. Such specificity is needed because of the particular situation of technology education as a relative newcomer in the curriculum still struggling with its public image and to some extent also with its identity, although the latter issue has been addressed extensively and much has been gained.

In technology education, we are still taken too much by surprise when policy makers decide in ways we cannot understand because "we had told them so clearly ...". Greater insight is needed into the mechanisms of educational policy making, not in general, but for very particular cases, such as technology education. This requires a very different type of study than we have seen published in our journals to date. It is still a rare type of research, even though the need for it is evident when considering the survival of technology education.

## The Epistemic Basis

Another important area for future technology education research is the epistemic basis for the subject. Recently, in the UK, the subject Design and Technology was critiqued for lacking such a basis, and as a result the position of Design and Technology in the curriculum was questioned. This poses a challenge for technology education research world-wide. In the first instance, we need to identify the epistemic basis of the subject; in other words, we have to find an agreed answer to the question: What are the fundamental concepts, laws and principles, in technology, that put the subject on an equal level with science and mathematics education? Some work has already been done on this and we are fortunate to have several studies that resulted in more or less the same list of basic concepts (Custer et al. 2010; Rossouw et al. 2011).

More is needed, though, than this list. We also need to know how the theory plays out in education. How can these concepts be taught and learnt? The suggestion has been taken from developments in science education that contexts should play a vital role in the learning process. Contexts, then, ought to be more than occasional examples to illustrate theory, but social practices that make sense to pupils and can only be participated in meaningfully with a proper mastery of certain concepts. Design activities are a candidate for such practices, but we still have little insight into how they stimulate conceptual learning. Some studies suggest they do, but other studies show no effect. No doubt, the problem is not in the individual studies, but in the fact that they are not easily comparable and therefore do not yet 'add up'. Additionally, they are generally not of a design-based research methodology so that the knowledge gained about the effect of design or other classroom interventions on conceptual learning was not based on a systematically optimised situation, but on a random existing one. Here, much work is still to be done.

Conceptual learning in design and technology is particularly challenging because of its abstract nature: we never see these concepts in practice. For example, what we see are cars, mobile phones, computers, buildings; not 'systems'. It takes time for us to learn that all the artefacts we see around us have certain characteristics in common that we use to understand a concept called 'system'. Even when we know what a 'system' is, it is sometimes hard to recognise it in a concrete artefact, because in each and every artefact this concept takes a somewhat different shape because the characteristics that are common for all systems are then mixed with characteristics that are specific to the car, the mobile phone, the computer or the building. Learning concepts therefore requires that we learn to separate the common characteristics from the context-specific characteristics. One can illustrate this by thinking about chameleons. The first time we meet one, it sits near the water and is blue. So we develop the idea that the chameleon is a blue animal with a long tongue. The next one we see is in the grass. We do not recognise it as a chameleon as it is not blue—yet it does have the long tongue. Then we see a third one on a red tiled roof and it is red, but has the same long tongue as the previous two had. Gradually we start realising that it is the tongue rather than the color of the body that makes the chameleon a chameleon. Once we know this, we recognise more easily that the grey animal sitting on the asphalt road with the long tongue is again a chameleon. Educational research can investigate how this learning process can be best organised and supported.

Teachers are a crucial aspect in realising high quality technology education. In science education, the Pedagogical Content Knowledge (PCK) of teachers has become a focus for study. Although the concept of PCK is still somewhat fuzzy, it has quickly grown in popularity. Broadly speaking, it is the knowledge that teachers need to have in order to be able to effectively teach specific topics. It is a very personal knowledge that teachers develop in the course of their education and teaching practice. Often it is associated with a European term for knowledge that is specific for one subject—vakdidaktiek in Dutch, Fackdidaktik in German, etc.

Preliminary studies investigating the nature of PCK in technology education have been conducted, but there is still a lot of work to be done. Measuring teachers' PCK is desirable, not in the least for teacher education programmes to evaluate the effects of the programme, but we still do not know how to measure the various components of PCK in technology education—or indeed in science education, where much research has been conducted. While much can no doubt be learnt from science education, technology education has many aspects that differ from science education. Research by Williams and Lockley (2012) indicated that the different nature of science and technological knowledge also rendered elements of PCK inapplicable to technology teachers. For some other elements, insights from science education can probably be transferred to technology education, with modifications regarding specific characteristics of technological concepts (such as normativity).

## Research on STEM

Another issue related to the survival of technology education in the curriculum is the relation with other subjects, and in particular with science and mathematics education. In several countries the acronym STEM is used to express the desire to interrelate science, technology, (pre-university) engineering and mathematics education. This can be done at several levels, ranging from accidental shared projects to a fully integrated school subject called STEM. As indicated by Cathy Buntting and Alister Jones in Chap. 10 of this volume, there are numerous hurdles for true integration of the components in STEM. One is that it is not easy to define projects in which knowledge and skills from science, technology, engineering and mathematics are all essential for success. Often the emphasis is on one or two of the STEM disciplines and the involvement of the other subject areas is quickly recognised by pupils as artificial. In addition, the ways in which the methods of the disciplines interact (research, design, mathematical modelling) is a matter that needs further research; in fact, this is almost entirely un-researched territory.

Related to this is the need for research into how various skills should cooperate in STEM. Creativity is definitely one of the skills that features in this spectrum—one that in fact is part not only of technology and engineering, but also science and to some extent mathematics. Studies were carried out decades ago into the way creativity develops in children (Torrence 1972), but little research is available about the way creativity can be developed by 'rich projects' in which design activities are combined with research activities and mathematical modelling (see Chap. 9 of this volume, by David Spendlove, for more on creativity in design). The need for and development of reasoning skills also needs further investigation. Here it is necessary to distinguish between cause-effect reasoning (can I reason back and forth between causes and effects?) and means-ends reasoning (can I reason back and forth between means and ends?). Both are important for science and mathematics as well as for technology and engineering. This is why integrated STEM projects should be a good vehicle for simultaneously developing skills in both reasoning types.

Although I have thus far grouped 'technology and engineering' together, the two components of the term can be distinguished from each other. Technology is the umbrella term for both the development and production of artefacts, systems and processes by engineers and technicians, and the use of such artefacts, systems and processes by engineers, technicians and users in society. Engineering is a narrower term that focuses on design and making by professionals who have been educated for such specialised work (engineers). Engineering is also characterised by certain concepts that can be absent in technology in the broader sense (Katehi et al. 2009). For example, engineering is highly quantitative in nature. A lot of design work in schools is qualitative and does not entail making calculations about constructions. For engineers, however, making calculations is a necessary part of their work. Another difference between technology and engineering is modelling, which tends to be far more prominent in engineering than in technology curriculum. In fact, modelling seems to be almost absent in much of technology education practice. Even if it is present, it remains implicit and the nature of models is not discussed explicitly. One exception is the New Zealand Curriculum, which includes technological modelling as an explicit component of technological knowledge. A third engineering characteristic is the intensive use of knowledge from natural sciences. This, too, is often lacking in current technology education practice. In other words, it can be concluded that the 'E' in STEM is still largely absent in current practice and both research *and* development are needed in the future to strive towards addressing this. This, too, has everything to do with the relevance and survival of technology education. Engineering is socially respected, and if technology education is not related to engineering in a visible way, it will miss this opportunity of being recognised as being socially relevant. As we have currently only begun acknowledging that basic engineering concepts should be taught, we are still some way from knowing how to teach them effectively. Research therefore has the potential of offering important support to developing the E in STEM.

A closer link with science education is not only important in the context of the STEM ideal in teaching practice, but also for research. Science education has a well-established tradition of educational research, from which technology education has a lot to gain. Numerous studies into both pupils' naïve or pre-concepts have been done in science education, while this research topic is still in its infancy in technology education. Instrument development is a crucial issue here. In science education research a lot of expertise has been developed and it would be a waste to start from the beginning in technology education. Additionally, strategies to identify or 'measure' initial ideas in more qualitative ways have been developed in science education research. Often the term misconceptions was used in this context. I would, however, want to argue against the use of that term, in line with current developments in science education. The pre-concepts that we have intuitively developed through practical experiences are often effective as long as their application is restricted to those situations in which we developed them. For example, in practice, we do not encounter situations without friction. Therefore it makes sense to develop the view that objects on which no force is acting will come to a standstill. It is only when we are confronted with frictionless contexts—which will typically happen only in a classroom experiment—that the application of the Aristotelian concept of force fails. The pre-concept then appears to be wrong, and we need to adapt our ideas. Still, for everyday-life situations, our old notion will continue to function satisfactorily and we will fall back on it even after we have learnt the scientifically correct concept. This is what can be called the difference between 'street image' and 'school image'. The phenomenon is well known in science education, but is worthy of further investigation in technology education.

## User-Orientation

In the UK, the Design Council has presented 'A new vision for Design & Technology in schools' (see http://www.designcouncil.org.uk/our-work/Insight/Education-and-skills/Design–Technology-in-schools/). Although this new vision has elements that are UK-specific, much of what is in this new view applies to other countries. Therefore, it would be profitable for technology education research to take into account certain elements of this vision in developing an agenda for future research.

Aspects promoted by the Design Council include:

- stimulating design literacy by not only focusing on the designer perspective but also the user perspective,
- enhancing the links between arts, science and business in the design capacity,
- enhancing pupils' understanding of user-centred design approaches,
- including up-to-date design technologies,

- bringing design activities in a higher theoretical level, and
- enhancing relations with industry.

Some of these 'new directions' are not as new as they are presented. However, one that is of particular interest because it seems to be undervalued in current technology education practice is the user perspective, both in the design literacy that is aimed for and in the design methods that are used. Research also has a blind spot here. To date, most research has taken for granted that design activities are there to develop design skills. Most pupils, however, will not continue their education to become a designer or engineer. They will, however, all become citizens who will use technologies. Almost no research has been done regarding how design and other activities can be developed and executed in order to stimulate user literacy. The users' perspective is also often missing in existing design activities in that pupils are mostly challenged to make the design such that it fulfils its 'technical' function. For example, the best mousetrap car design is the one that travels the longest distance. Often no user is defined and pupils do not need to bother explicitly about a user's perspective. Research has reflected this absence of emphasis, and few studies focus on how a user-focused approach in design projects can be developed with pupils. David Spendlove also raises the issue of user-orientation in Chap. 9 of this volume.

## The Use of Media

The use of media has long been an area of interest in technology education research. There are, however, currently important reasons for affording this area a more prominent place in technology education. The emergence of gaming and social media have had an enormous impact on the lives of young people, and their interactions with such media have become a substantial part of their daily activities. To date, technology education has been very much bound to school classes, perhaps extended with ad-hoc visits to museums or industrial organisations.

While school continues to be a place where pupils spend much of their time, an increasing part of their lives has become virtual. They love gaming, become inhabitants of virtual worlds such as Second Life, are plugged into various social networking sites, and many write blogs. This has at least two implications for technology education. First, a new dimension should be added to technological literacy, as it has become evident that appropriate use of these media by young people is by no means to be taken for granted. For example, reports have been published that show that pupils perform worse in education because they are too preoccupied with the use of games (e.g., Rehbein et al. 2010). An addictive effect of such media has also been reported. This means that contemporary technological literacy also entails the knowledge, skills and attitudes to give social media an appropriate place in one's life. Educational research can support the development of new pedagogies and materials that help learners acquire such dimension in technological literacy.

The second implication is the use of such media to teach about technology. While the use of media in education is the domain of educational technology rather than technology education, when we focus on the use of such media to teach specifically about technology, it becomes part of technology education. For this reason it should form part of technology education research when it comes to developing insights into how digital technologies can be used for educational purposes.

## Teacher Education

I will be short on research about teacher education, as the research topics addressed above all have implications for research in teacher education. If teachers are to be given a more active role in research, they need to be prepared for this. It is by no means obvious how to do this as programmes are often already overloaded and it is not likely to be possible in the short time available to prepare teachers adequately for doing research—or even to interact intellectually with it. The small research studies student teachers are required to conduct tend to be fairly superficial and don't meet the standards of academic research.

Research is needed to identify feasible ways of acquainting pre-service teachers with research methodologies in ways that fit within the dimensions of a teacher education programme. Two aims seem to be realistic: preparing teachers for being good partners with researchers while still primarily being teachers, and giving teachers some experience in research by having them assist in faculty members' research. In the latter case, they do not need to do a full research study themselves but only be part of a certain phase of a research study, and learn about the other phases by listening to the researchers' stories. In such a master-apprentice situation, future teachers can get a taste of what research is without the burden of a complete mini-research project that they are individually responsible for carrying out. Research in teacher education could show how this can be given shape in an effective way. Those who are already practicing teachers who want to become involved in research will need to be provided with in-service activities to acquaint them with current research methodologies and the way these can be applied in technology education research.

## The Effect of Research on Policy and Practice

How can we know the effect of research on teaching practice and policy? This issue is also addressed by Kendall Starkweather in Chap. 13 of this book. Often, the relationship between research and policy making is indirect. What may happen is that teacher educators get acquainted with outcomes of educational research and incorporate the outcomes in their teacher education programmes. They often work alongside researcher colleagues, and in many cases carry out research themselves (refer to my earlier plea for this!). The result of a changed teacher education is a new generation of teachers who have knowledge that has originated from research projects. There is therefore a time gap of at least one generation in this mechanism for research to affect practice. This also makes it difficult to trace back the educational changes to the original research efforts that led to them. A challenge for the future is to make the relation more explicit. This can be done, for instance, by offering the research-based insights taught in teacher education in such a way that the future teachers realise that it was research that produced these insights and not the intuition and experience of the teacher educators. Teachers should also be taught to read research articles so that they are equipped to acquaint themselves with future findings from educational research. It should become 'second nature' for teachers to monitor the research field of their teaching subject. Close cooperation between researchers and the teacher educators can stimulate this.

Another way in which research can impact on the practice of technology education is through the emergence of international contacts. In particular, research-oriented conferences and workshops have contributed to increasing international cooperation in technology education. For example, the meeting of researchers at international research-oriented conferences facilitated the spreading of knowledge about ways in which different countries were approach problems in developing and introducing technology education, providing a source of inspiration for curriculum developers and teacher educators in other countries. The PATT conferences are an example of conferences with such an effect. Although they originally focused exclusively on attitude measurement, the scope soon widened to include other aspects of technology education. The research-oriented nature of the early PATT conferences has been maintained throughout the existence of the series of conferences. Thus, reported and discussed research efforts have had a certain impact on the internationalisation of technology education and made it possible for strengths from different countries to be built on, and experienced weaknesses to be avoided. This is a form of effectiveness that, like the previous mechanism (teacher education influenced by research), is seldom documented, and it is again difficult to assess precise ways in which research has affected practice.

## Concluding Remarks

In this chapter I have sketched future directions for technology education research. I have suggested an agenda for technology education research in which design-based studies will become a more prominent type of study. In addition, research questions will be chosen that relate more directly to what interests and is valued by teachers, particularly the effect of new and innovative learning materials. I also made a plea for research into the process of policy making in technology education, the challenges that STEM education poses, the user-dimension of technological literacy, and the use of new media in technology education. Teachers should play a more active role than currently in carrying out this research agenda.

Whether or not this road can be travelled will depend on the investments that are made. In this regard, there is some concern. Several institutes that have been strong in technology education research have given up their efforts. In Germany, for instance, there were several universities with robust technology education research, but many of them are no longer active in this field. In the UK, several institutes have also closed their research programmes. In the former Eastern-European countries almost no research in technology education survived the political changes of the 1980s/1990s. On the other hand, in countries like New Zealand, there has been an increase in research activities in the twenty-first century.

It is important that universities invest in technology education research. Unfortunately, there is a vicious circle here. Technology education research will only be valued when technology education itself is valued—but to provide justification for the appreciation of technology education, relevant research outputs are needed. It is still possible that technology education will go through a phase in which its very survival will depend on the efforts of a few countries that perform extremely well in technology education research and thus provide the ammunition for others to defend both technology education itself and its supporting research to their governments and policy makers. The extent to which researchers will be able to realise a better link with educational practice and enhance their understanding of policy processes will, in my view, determine the future of technology education. There is every reason to work hard on this, as our case is definitely worth it.

## References


Broekkamp, H., & van Hout-Wolters, B. H. A. M. (2007). The gap between educational research and practice: A literature review. *Educational Research and Evaluation, 13*, 303–320.

Custer, R. L., Daugherty, J. L., & Meyer, J. P. (2010). Formulating a concept base for secondary level engineering: A review and synthesis. *Journal of Technology Education, 22*(1), 4–21.

de Vries, M. J. (2003). Editorial. *International Journal of Technology and Design Education, 13*, 199–205.

Foster, W. T. (1992). Topics and methods of recent graduate student research in Industrial Education and related fields. *Journal of Industrial Teacher Education, 30*(1), 59–72.

Garmire, E., & Pearson, G. (2006). *Tech tally. Approaches to assessing technological literacy*. Washington, DC: National Academies Press.

Jones, A., & de Vries, M. J. (Eds.). (2009). *International handbook of research and development in technology education*. Rotterdam/Taipei: Sense.

Katehi, L., Pearson, G., & Feder, M. (2009). *Engineering in K–12 education: Understanding the status and improving the prospects*. Washington, DC: National Academies Press.

Kimbell, R., & Stables, K. (2007). *Researching design learning. Issues and findings from two decades of research and development*. Dordrecht: Springer.

Petrina, S. (1998). The politics of research in technology education: A critical content and discourse analysis of the Journal of Technology Education, Volumes 1-8. *Journal of Technology Education, 10*, 27–57.

Rehbein, F., Kleimann, M., & Mößle, T. (2010). Prevalence and risk factors of video game dependency in adolescence: Results of a German nationwide survey. *Cyberpsychology, Behavior and Social Networking, 13*, 269–277.

Rossouw, A., Hacker, M., & de Vries, M. J. (2011). Concepts and contexts in engineering and technology education: An international and interdisciplinary Delphi study. *International Journal of Technology and Design Education, 4*, 409–424.

Sjøberg, S., & Schreiner, C. (2010). *The ROSE project: An overview and key findings*. Oslo: University of Oslo.

Skogh, I.-B., & Gumaelius, L. (2012). Technology teachers as researchers: The TUFF experience. In H. Middleton (Ed.), *Explorations of best practice in Technology, Design & Engineering Education* (Vol. 2, pp. 118–127). Brisbane: Griffith Institute for Educational Research.

The Design-Based Research Collective. (2003). Design-based research: An emerging paradigm for educational inquiry. *Educational Researcher, 32*, 5–8.

Torrence, E. P. (1972). Can we teach children to think creatively. *The Journal of Creative Behavior, 6*, 114–143.

Wicklein, R., & Hill, R. B. (1996). Navigating the straits with research or opinion? Setting the course for technology education. *International Journal of Technology and Design Education, 6*, 31–43.

Williams, P. J., & Lockley, J. (2012, July). *An analysis of PCK to elaborate the difference between scientific and technological knowledge*. Paper presented at the Pupils Attitude Toward Technology (PATT) 26 Conference: Technology Education for the 21st Century. Stockholm.

Zuga, K. F. (1997). An analysis of technology education in the United States based upon an historical overview and review of contemporary curriculum research. *International Journal of Technology and Design Education, 7*, 203–217.

# Chapter 15
# Much Remains to Be Done

**Alister Jones, Cathy Buntting, and P John Williams**

This book took as its starting point the assumption that, even in an ever-changing environment, some things stay the same. Education policy needs to coherently connect government and community aims across all levels of education. Education systems need to be sufficiently robust and flexible to deliver effective learning and assessment opportunities to diverse students. The curriculum—which reflects government policy and the systems within which it operates—should meet the needs of students and the wider community. But while these principles are unchanged, what varies is the detail: how (and whether) policy connections are established and maintained, features of the education system that are needed for it to be both robust and flexible, and what the intended and implemented curriculum looks like.

Technology education is new enough as a school subject that many readers will remember versions of its craft-type predecessors. The current iteration of the subject, by and large, embraces learning *in* and *about* technology and developing skills in design, creativity, problem solving, systems understanding and critique, in addition to specific technical skills. 'Technological literacy' has become embedded within curriculum and political rhetoric, although references to 'technology for all' often sit alongside vocational purposes for the subject. This shift from craft subjects has been accompanied, in many cases, with an increased focus on the

A. Jones (✉)
Deputy Vice-Chancellor, University of Waikato, Hamilton, New Zealand
e-mail: a.jones@waikato.ac.nz

C. Buntting
Faculty of Education, University of Waikato, Hamilton, New Zealand
e-mail: buntting@waikato.ac.nz

PJ. Williams
Technology, Environmental, Mathematics and Science Education Research Centre, University of Waikato, Hamilton, New Zealand
e-mail: jwilliam@waikato.ac.nz



PJ. Williams et al. (eds.), *The Future of Technology Education*, Contemporary Issues in Technology Education, DOI 10.1007/978-981-287-170-1_15

values-oriented activity of design and technology, including values associated with environmental, economic and social sustainability and responsibility.

The history of the development of technology education has been documented by many, some making international generalisations while others have emphasised unique national or regional developments. There can be no doubt that significant advances have been made in both the breadth and depth of technology education since its introduction as a distinct learning area. In spite of this ongoing development, however, the subject remains susceptible to the vagaries of political whims and system disconnects.

For example, in England there is a national curriculum for Design and Technology, design awareness is a strong component of society, and the professional association has broad membership and is actively engaged in support and promotional work. In spite of this, Design and Technology has recently been under threat of being removed as a compulsory part of the curriculum. In the USA there is significant diversity in technology education across the country but a strong professional association. However, the drive for general and political recognition, which has resulted in a move to STEM and engineering, may result in the regeneration of technology education in a new form. The impact of the recent inclusion of technology and engineering standards within the science standards remains to be seen. Within Australia, recent development of the national curriculum has meant that Technology Education is now integrated with ICT, though it is one of the last tier of subjects to be developed. New Zealand's national curriculum for Technology Education is research based. However, despite significant government-funded professional development and a strong professional association, there continues to be a significant gap between curriculum documentation and practice, and the sub-culture associated with former 'wood work', 'metal work' and 'home economics' subjects has deep roots. South Africa has a national curriculum but a weak professional association, little standardised teacher training and professional development specifically targeting Technology Education, and few resources. It seems that teacher enthusiasm for the subject is currently what sustains it in the curriculum. Countries in the Global South tend to have fragile education systems and significant challenges with respect to overpopulation, struggling economies, high unemployment and low literacy. Here, the place of technology education as a vocationally relevant subject may be particularly pertinent.

Dilemmas between vocational and general approaches to technology education remain in many educational jurisdictions. The existence of a national curriculum presumes a general rationale, namely that the subject is important for all students to study regardless of the type of work they will do when they leave school. However, at the school level, particularly where the provision of vocational education has been historically important and where the same teachers teach both approaches, there is often confusion about the goals, attributes, pedagogy and assessment.

It is clear, however, that a robust international research agenda will not be enough to ensure that technology education is valued at all levels, or that informed and effective practice is developed. As evidenced throughout this book, research in technology education needs to be relevant and accessible to teachers. Involving teachers more closely in research practices is one step towards achieving closer research-practice links. Another is closer relationships between researchers and teacher educators, both for pre-service and in-service education. Although current international accountability measures for researchers creates tensions between publishing in academic versus teacher-accessible forums, we need technology education researchers who are both committed to publishing in teacher journals, and active in teacher professional organisations.

Professional organisations provide professional support to teachers, as well as playing key roles with respect to advocacy and political persuasion, and building brand awareness. However, the changing social and technological context of education means that such professional organisations will need to actively adapt if they are to continue to meet the needs of teachers and the profession. Ease of access to international thinking via social media and other Web 2.0 technologies offers tantalising opportunities for examples of effective practice to cross between education systems. We also need to publicly celebrate students' successes and achievements in technology education in order to raise the profile of the subject—successes that come not only through elite competitions for a few, but from measurements that educational and political systems value.

Work to develop valid, authentic ways to assess students' learning in technology education has been enormously strengthened by the undertakings of research groups such as the Technology Education Research Unit at Goldsmiths, University of London. However, we are still a long way from embedding authentic assessments—such as are provided by e-portfolios—in high-stakes education contexts. International assessments like TIMSS and PISA also need to be carefully interpreted, although there is unfortunately a long history of ill-informed 'knee jerk' reactions to lower-than-expected rankings. In other words, high TIMSS and PISA achievement should not be pursued as an aim in itself. Rather, we should expect—and provide evidence—that good technology education results in positive outcomes in studies that have been more broadly conceived.

Aligning assessment with the purpose(s) of technology education is a critical but difficult undertaking. It relies on clearly identifying the intended learning outcomes, and access to valid formative and summative strategies for assessing the learning. Extremely high expectations are placed on teachers, who are required to interpret curriculum intentions and translate these into engaging, effective learning opportunities for diverse students. Pedagogical practice can be enhanced through formal and informal professional learning opportunities, as well as ongoing reflection. However, support for ongoing professional learning of teachers—in specific subject areas, let alone to support intentional cross-curricular learning—is in many jurisdictions not sufficiently funded, or even at times supported by teachers themselves.

As the educational research community continues to grapple with the needs and aspirations of students now that we are well into the twenty-first century, there continues to be too little real evidence of the crucial role that technology education can play in developing citizens who will contribute to and engage with the significant issues that our world faces. Longitudinal studies are needed to investigate the key understandings and skills, and the personal attributes, that are developed through effective technology education, and the impact that this development has on students' future lives.

The growing emphasis on STEM, and even STEAM (science-technology-engineering-arts-mathematics) in some places, signals a need for technology education to develop a space within cross-curricular learning where its core values can be embedded and propagated. What will it take to create a future where teachers of all subjects recognise the contribution that can be made by technology learning, and where they seek ways to support such learning? While science continues to dominate the learning sought under the STEM mantra, it is technological applications and their critique that are of far greater importance when responding to current and future challenges faced by societies and the world.

An enormous amount has been accomplished over the last three decades or so in establishing technology education as a distinct school discipline. However, global changes in terms of knowledge creation and distribution mean that if technology learning is to be truly relevant and valued, there remains much to be done. We hope that this book contributes usefully to the debate.

# Index



 
PJ. Williams et al. (eds.), *The Future of Technology Education*, Contemporary Issues in Technology Education, DOI 10.1007/978-981-287-170-1

**D**

MIX
Papier aus verantwortungsvollen Quellen
Paper from responsible sources
FSC® C105338

If you have any concerns about our products,
you can contact us on
**ProductSafety@springernature.com**

In case Publisher is established outside the EU,
the EU authorized representative is:
**Springer Nature Customer Service Center GmbH**
**Europaplatz 3, 69115 Heidelberg, Germany**

Printed by Libri Plureos GmbH
in Hamburg, Germany